The Impact of the Laboratory and Technology on Learning and Teaching Science K-16

A volume in
Research in Science Education

Series Editors:
Dennis W. Sunal, *University of Alabama*
Emmett L. Wright, *Kansas State University*

Research in Science Education

Dennis W. Sunal and Emmett L. Wright, Series Editors

Reform in Undergraduate Science Teaching for the 21st Century (2003)
edited by Dennis W. Sunal, Emmett L. Wright, and Jeanelle Bland

The Impact of State and National Standards on K-12 Science Teaching (2006)
edited by Dennis W. Sunal and Emmett L. Wright

The Impact of the Laboratory and Technology on Learning and Teaching Science K-16

edited by

Dennis W. Sunal
University of Alabama

Emmett L. Wright
Kansas State University

and

Cheryl Sundberg
University of Alabama

Information Age Publishing, Inc.
Charlotte, North Carolina • www.infoagepub.com

Library of Congress Cataloging-in-Publication Data

The impact of the laboratory and technology on learning and teaching science K-16 / edited by Dennis W. Sunal, Emmett L. Wright and Cheryl Sundberg.
    p. cm. -- (Research in science education)
 Includes bibliographical references.
 ISBN-13: 978-1-59311-744-3 (pbk.)
 ISBN-13: 978-1-59311-745-0 (hardcover)
 1. Laboratories--Technique--Study and teaching 2. Science--Study and teaching. I. Sunal, Dennis W. II. Wright, Emmett. III. Sundberg, Cheryl.
 Q183.A1.I67 2008
 507.1--dc22

                                        2007046229

ISBN 13: 978-1-59311-744-3 (pbk.)
         978-1-59311-745-0 (hardcover)

This volume is jointly sponsored by the Office of Research on Teaching in the Disciplines at the University of Alabama.

Printed in the United States of America

CONTENTS

PREFACE TO THE SERIES

Science education as a professional field has been changing rapidly over the past 2 decades. Scholars, administrators, practitioners, and students preparing to become teachers of science find it difficult to keep abreast of relevant and applicable knowledge concerning research, leadership, policy, curricula, teaching, and learning that improve science instruction and student science learning. The literature available reports a broad spectrum of diverse science education research, making the search for valid materials on a specific area time-consuming and tedious.

Science education professionals at all levels need to be able to access a comprehensive, timely, and valid source of knowledge about the emerging body of research, theory, policy, and practice in their fields. This body of knowledge would inform researchers about emerging trends in research, research procedures, and technological assistance in key areas of science education. It would inform policymakers in need of information about specific areas in which they make key decisions. It would also help practitioners and students become aware of current research knowledge, policy, and best practice in their fields.

For these reasons, the goal of the book series, *Research in Science Education*, is to provide a comprehensive view of current and emerging knowledge, research strategies, and policy in specific professional fields of science education. This series presents currently unavailable, or difficult to gather, materials from a variety of viewpoints and sources in a usable and organized format.

Each volume in the series presents a juried, scholarly, and accessible review of research, theory, and/or policy in a specific field of science education, K-16. Topics covered in each volume are determined by current issues and trends, as well as generative themes related to up-to-date research findings and accepted theory. Published volumes will include empirical studies, policy analysis, literature reviews, and positing of theoretical and conceptual bases.

PREFACE

The Impact of the Laboratory and Technology on K-12 Science Learning and Teaching examines the development, use, and influence of active laboratory experiences and the integration of technology in science teaching. This examination involves the viewpoints of policymakers, researchers, and teachers that are expressed through research involving original documents, interviews, analysis and synthesis of the literature, case studies, narrative studies, observations of teachers and students, and assessment of student learning outcomes. Volume 3 of the series, *Research in Science Education*, addresses the needs of various constituencies including teachers, administrators, higher education science and science education faculty, policymakers, governmental and professional agencies, and the business community.

The guiding theme of this volume is the role of practical laboratory work and the use of technology in science learning and teaching, K-16. The volume investigates issues and concerns related to this theme through various perspectives addressing design, research, professional practice, and evaluation. Beginning with definitions, the historical evolution and policy guiding these learning experiences are explored from several viewpoints. Effective design and implementation of laboratory work and technology experiences is examined for elementary and high school classrooms as well as for undergraduate science laboratories, informal settings, and science education courses and programs. In general, recent research provides evidence that students do benefit from inquiry-based laboratory and technology experiences that are integrated with classroom science curricula. The impact and status of laboratory and technology experiences is addressed by exploring specific strategies in a variety of scientific fields and courses. The chapters outline and describe in detail research-based best practices for a variety of settings.

Of particular interest and use, for effective planning of science instruction, are the definitions of (1) laboratory work—as learning activities incorporating hands-on and minds-on interactive strategies involving inquiry experiences integrated with instruction in everyday science teaching, taking place in a classroom, laboratory, or in informal settings and (2) technology learning experiences—as evidence gathering or communicating activities involving any tool, piece of equipment or device, electronic or mechanical, that can be used to help students accomplish specified science learning goals These chapters provide an evidence-based view of the current status and future of the integration of active learning through laboratory and technology experiences and of the work that still needs to be done to meet our expectations for improving science learning and teaching.

ACKNOWLEDGMENTS

This third volume in the series *Research in Science Education* was made possible by our authors and colleagues, who gave priority in their professional lives to conduct investigations, write manuscripts, submit their work to the scrutiny of others, and persist through many revisions. Their professional experiences and expertise makes this volume possible. Our contacts with the authors developed through our long term membership in professional organizations and activities with numerous individuals in education institutions. Therefore, we wish to acknowledge these organizations and individuals for providing us with a forum to meet, interact, disseminate, and form professional collaboration communities involving individuals with an interest in improving teaching and learning in science. We especially thank the National Association for Research in Science Teaching, National Science Foundation, Association of Educators of Teachers of Science, American Association for the Advancement of Science, National Science Teachers Association, American Educational Research Association, and the National Aeronautics and Space Administration NOVA Program and Network of university teams involved in reform. Special recognition is given to the students and teachers who were concerned enough to take part in the investigations and contribute their thoughts to our discussions.

We also acknowledge our graduate students at various institutions who in addition to completing their own research and course work undertook tasks that allowed this volume to come to completion. Special thanks go to Kim Mantle and Michael Hallman at the University of Alabama.

Dennis W. Sunal
Emmett L. Wright
Cheryl Sundbsrg
June, 2007

CHAPTER 1

THE IMPORTANCE OF LABORATORY WORK AND TECHNOLOGY IN SCIENCE TEACHING

Dennis W. Sunal, Cynthia S. Sunal, Cheryl Sundberg, and Emmett L. Wright

The use of the science laboratory and technology has been a part of science teaching in the United States for more than 120 years. Teachers are aware of, and reflect on, the use of both almost on a daily basis. With this long history, the chapter focuses on investigating the question "What is the importance of laboratory work and technology integration in science teaching, K-16?" The viewpoints of policy documents and stakeholders involved in planning for or using the science laboratory and technology provide the focus of the chapter. An investigation of teachers' perceptions and expectations of the science laboratory and technology demonstrated that, while the majority was aware of the importance, sizable percentages lacked knowledge of purpose and use of the science laboratory and technology in daily classroom practice. Finally, the importance of laboratory work and technology experiences in teaching science requires knowledge and skills to analyze and adapt procedures. The analysis of inquiry potential instrument is described with an example of its use. The purpose of the instrument is as a guide and

The Impact of the Laboratory and Technology on Learning and Teaching Science K-16, pp. 1–28

an aid for teachers in gathering objective data when making critical judgments on planning, use, and adaptation of existing laboratory materials based on the goals and recognized principles of good science education.

INTRODUCTION

The use of the science laboratory and technology has been a part of science teaching in the United States since at least the middle of the 1800s. It is difficult to separate the coevolution of laboratory work and technology experiences in American science classrooms. Asking students to investigate phenomena associated with water as it is being heated to boiling by measuring and graphing the temperature with a thermometer, or with a sensor temperature probe, is an example of both the use of laboratory work and technology in teaching.

Teachers give numerous reasons for using laboratories and technology in teaching science. Some reasons include interest, curiosity, higher order thinking, and learning science concepts. An example of one teacher's understanding related to learning lesson concepts is, "I am using rotational stations for science laboratory experiences, because last year I did experiments as demonstrations and everyone could not see and some students did not get concept" (fifth grade teacher, Shelby County, Alabama). Another teacher was concerned about unique learning outcomes that result from laboratory work.

> Doing laboratory activities with middle grade students is imperative because of the extreme curiosity of this age student. Weekly hands-on activities and frequent inquiry-based experiments provided my students with opportunities to be independent thinkers and problem solvers not only in the science classroom but in their own everyday lives. (fifth grade teacher, Glenda Ogletree, Shelby County, Alabama)

Determining the importance of the laboratory work and the use of technology to facilitate science learning and teaching, however, requires a review of policy, research, and practice. Of concern are issues and events surrounding the development, status, influence, and future of teachers' use of laboratory and technology experiences in science teaching. The viewpoints of researchers, policymakers, teachers, and students have been expressed frequently over the past 30 years through documents, interviews, analyses of the literature, case studies, ethnographic and narrative studies, observations of classrooms, and assessment of student learning outcomes.

In schools today, laboratory work is part of science lessons about 1 day a week in both elementary and secondary schools (Singer, Hilton, & Schweingruber, 2005). Elementary teachers report spending an average of 24 minutes a day in Grades K-3, and 31 minutes a day in Grades 4-6 in teaching science with about 25% of the time in laboratory work (Hudson, McMahon, & Overstreet, 2000). Secondary science teachers report spending 45 minutes per day in teaching science with 25% of the time in laboratory work. More recent data from a national study show a decrease in elementary schools of an average of 15 minutes per week for science since the enactment of the No Child Left Behind (NCLB) act in 2001 and an increased focus on English language arts (Center for Education Policy, 2007). The amounts of laboratory and technology experiences have also decreased. Variation exists with some students receiving little, if any, laboratory work, or any science at all, while others are engaged with science learning on most school days. These outcomes create a two tier educational system for future career paths.

The use of laboratories in science teaching gradually started as demonstration activities in universities from the 1840s to the 1880s, evolving into more student engagement. Early adopters were at Yale, Rensselaer Institute, and John Hopkins University.

In the elementary school, science laboratory experiences and the development of laboratory skills started in the 1850s. The introduction of student engagement with objects and nature was influenced by European educators such as Johann Pestalozzi (1746-1847), Friedrich Froebel (1782-1852), and Herbert Spencer (1829-1903). The German Pestalozzian "object teaching" became known in America as the Oswego, New York "Method." Its purpose was to train observation, description, and memorization of animate and inanimate objects as preparation for studying science in upper grades. This curriculum was very fragmented and the emphasis was on description with the interpretation and understanding of events and phenomena neglected. Influenced by Herbert Spencer and the rise of popular interest in science technology in the 1870s, science was pushed as a field of study in elementary schools. The first organized science curriculum for elementary schools began in the 1870s. By the 1890s resurgence in laboratory work accompanied the nature study movement, whose purpose was to help children learn about and appreciate their environment. Criticism of this movement focused on the fragmented identification and isolated bits of data with overemphasis on sentimental, emotional, and aesthetic explanations.

High school science laboratory work began in the late 1800s as a prerequisite for students preparing for the university (Singer, Hilton, & Schweingruber, 2005). This reform was influenced by Edwin Hall (1855-1938), whose publication of a list of science labs to be completed spurred the need for students to enter college with substantial laboratory work. A report from the National Education Association in 1894 stating the need for laboratory work consolidated laboratory reform in science teaching (Rudolph, 2002).

From these early times to the present, views on the purpose and necessity of laboratory work have waned and strengthened (Singer, Hilton, & Schweingruber, 2005). As schools expanded dramatically from the 1890s through 1910 due to immigration, a decrease occurred in laboratory work because of the large amount of resources needed to serve the basic needs of rapidly increasing numbers of students. A greater focus on traditional content resulted.

With the advent of the new educational psychology movement led by William James (1842-1910), John Dewey (1859-1952) and others, student engagement in learning leading to science laboratory work began to increase in schools from 1910 through the early 1940s. More emphasis was made on practical content and student activity where the meaning of concepts was found in the process of experience. In the 1930s, under the influence of reports such as the National Society for the Study of Education, science curricula were based on major science principles and their application. During these 3 decades laboratory work for students K-16 became a regular part of traditional science teaching and learning.

After World War II an increase in students straining the educational system, the new goals of science education reflected a change toward life adjustment and purposes that met perceived student needs. Selection of science content was based upon personal and social criteria. This shift led to a decrease in the importance of laboratory work as part of the school day. Science teaching focused on the textbook and classroom lecture.

In the mid-1950s, spurred on by international competition with the Soviet Union, America again began to focus on science teaching through student engagement. Between the mid-1950s and 1977 a wide variety of federally-funded laboratory and technology oriented curricula were developed and implemented in schools K-12. Science teaching stressed learning through the discovery approach, hands-on learning, and used the structure and content of scientific disciplines for shaping the curricula. By 1977, over 60% of school districts had adopted at least one of the new national science curricula (Rudolph, 2002).

The "back to basics" emphasis in schools in the 1970s suggested that drill activities and an emphasis on lower-level understandings were important factors in improving science teaching. While researchers reported positive effects in fact level questions, national science achievement scores plummeted during this time. The time spent on drill and lower level understandings also was found to be detrimental to students' higher level understandings, use of thinking skills related to science, and experience with problem solving strategies used in science (Martens, 1992).

Under the influence of educational researchers such as Jerome Bruner (1915-present), Joseph Schwab (1908-1988), Piaget (1996-1980), and Vygotsky (1896-1934), the emphasis on laboratory work changed during the 1980s and 1990s. The emphasis moved from the hands-on approach of earlier decades to a laboratory approach involving hands-on and minds-on inquiry learning through investigations. Throughout the past several decades, influenced by frequent national reports providing new views on the nature and goals of science, the use of the laboratory again has increased. Some of the important reports are *A Nation at Risk: The Imperative for Educational Reform* (1983), *Science for All Americans* (American Association for the Advancement of Science [AAAS], 1989), *Benchmarks for Science Literacy: Project 2061* (AAAS, 1993), *Goals 2000: Educate America Act* (P.L. 103-227) (1994), and *National Science Education Standards* (National Research Council [NRC], 1996). While in today's schools the large majority of classrooms still reflect traditions dating from World War II, a number of classrooms involve students in regular laboratory work centered on inquiry learning as advocated in reports and research documents of the past 10 years. Descriptions of the characteristics of these inquiry based laboratory work and technology experiences follow in later sections of this chapter.

Some questions to be answered on the way toward understanding the importance of laboratory and technology experiences in teaching science include; "What practices are supported by policy documents today?," "What has been the response of today's teachers toward the role of practical laboratory work and technology in facilitating learning in the teaching of science?," and "What directions should laboratory work and technology experiences take in science teaching and learning in today's schools and universities?

POLICY AND THE IMPORTANCE OF LABORATORY WORK AND TECHNOLOGY IN TEACHING AND LEARNING SCIENCE

Some Definitions

Laboratory work and technology experience are defined here as incorporating hands-on and minds-on interactive learning strategies involving inquiry activities integrated with instruction in everyday science teaching, and taking place in a classroom, laboratory, or in informal settings. An inquiry strategy involves students in constructing their own meaning for science ideas they encounter, taking control of their own learning to get to the goal of a lesson through guidance from their teacher. During inquiry the student gives priority to evidence when responding to questions, constructing explanations, and justifying conclusions (NRC, 2000a). The integration of these experiences with the everyday teaching of key science curriculum concepts is a critical condition. Without integration into classroom science teaching and learning, the impact of the experiences is minimized and they become peripheral fun experiences that are not seen as important by students and make little positive impact on achievement (Sunal, Wright, & Sunal, 2008). The assessment of both laboratory and technology experiences is best seen as comparing the actual experiences of students' lab or technology work with the intended activities (Psillos & Niedderer, 2002). It is usually suggested that teachers assess work experience learning outcomes and also learning objectives in classroom settings.

Singer, Hilton, and Schweingruber (2005) defined laboratory work experiences as "opportunities for students to interact with the material world (or with data drawn from the material world), using the tools, data collection techniques, models, and theories of science" (p. ES-2). Laboratory work experiences result in student learning outcomes important in both the processes and content of science. The definition of laboratory work experiences also involves integrated sequences of instruction relating lectures, reading, discussion, and other learning experiences *with* students' laboratory experiences (Singer, Hilton, & Schweingruber, 2005). This teaching approach is contrasted against transmission pedagogy, direct instruction, and other methods that involve a focus on science facts or concept learning through memorization of definitions found in science textbooks. The definition of laboratory experiences also is contrasted with "traditional" laboratory experiences that are disconnected in time and sequence from other science instruction.

Educational technology in science is defined further in this chapter as including any tool, piece of equipment or device, electronic or mechanical, which can be used to help students accomplish specified science learning goals (NRC, 2000b). This is at odds with the more traditional view of technology as relating to computers or to software or to Internet usage. The definition used here lends itself to the understanding that technology is seen as a means or a tool to facilitate students' investigations in science. Students and science teachers use technology of many kinds, from spring balances to handheld computers, to support and promote thoughtful investigations of important questions in science. Teachers also use technology to represent scientific ideas and data in meaningful ways facilitative of students' learning.

Students' experiences with laboratories and technology are tools to create and facilitate students' science learning. In science education, K-16, they are not entities in themselves as they would be in the professional world of science and scientific research.

Documents Describing the Appropriate Use of the Laboratory and Technology

Several important documents are available that support and/or mandate laboratory work and technology use in teaching science. NCLB, the Elementary and Secondary Education Act passed in 2002, supports an emphasis on technology integration in K-12 science teaching (U.S. Department of Education, 2002). As an outcome of this act, education leaders at the state and local levels are expected to develop plans that effectively employ educational technology throughout the curriculum. The *National Science Education Standards* (NRC, 1996) defines appropriate science teaching as reflecting on the true nature of science, recognizing science as both a body of knowledge and a process to be developed using laboratory and technology experiences.

Laboratory work was the focus of a National Science Teacher's Association (NSTA) position paper early in 2007, *Integral Role of Laboratory Investigations in Science Instruction* (NSTA, 2007). This position paper addressing the needs of K-16 students added the criterion of developmentally appropriate and meaningful to the definition of laboratory science. This definition indicates that practice should avoid focusing on rote exercises using cookbook-type laboratory activities and disconnected proofs of phenomena partially related to the objectives of the science being taught in the science class at the time. Appropriate laboratory investigations have a clear purpose, focus on science processes first and content second, involve student reflection and discourse, and develop safe ways of working with materials and phenomena in laboratories and in daily life. NSTA's position insists that "For science to be taught properly and effectively, labs must be an integral part of the science curriculum" (NSTA, 2007, para. 3).

The NSTA paper describes an expectation that at the preschool and elementary level, all students regularly explore science phenomena in science "laboratories" several times a week. Student laboratory work includes asking questions and investigating science phenomena and reflecting on the results through discussion and communication of findings and conclusions. At the middle school and high school levels, science laboratory work must support classroom instruction through regular weekly laboratory experiences. These experiences build in complexity with age. Their focus is on collaboratively collecting, interpreting, and communicating evidence; using technology to support the work; reflecting on evidence through argumentation; and responding to and contributing ideas. Laboratory work for the college level student should be an integral part of every introductory science course. Rigorous inquiry laboratory work should help students develop the skills to independently and cooperatively use appropriate laboratory techniques, and to critique their own work as well as others.

The *NSTA Position Statement; Science/Technology/Society (STS): A New Effort for Providing Appropriate Science for All* (NSTA, 1990) supports the view that technology is best used as a tool to facilitate teachers' instructional strategies and students' learning in the science classroom. The NSTA position is that "The emerging research is clear in illustrating that science in an STS context results in students with more sophisticated concept mastery and ability to use process skills" (NSTA, 1990, para. 1). Technology, which forms part of STS, allows the teacher and students to focus on real-world problems instead of starting with a statement of a predetermined concept or process. STS allows for inductive and constructivist teaching and learning. Technology facilitates assisting students in identifying local or regional problems, planning for group activities that address these problems, and designing activities to assist students in resolving the problems investigated. The NSTA paper concludes by describing the scientifically and technologically literate person as appropriately using science and technology; using and valuing science and technology in solving problems; distinguishing between science and technology; understanding the strengths and limitations of science and technology for advancing human welfare; and considering political, economic, and ethical aspects of science and technology as they relate to personal and global issues.

More recent discussion of technology involves its impact on communication of the Internet, data acquisition systems, video gaming, blogs, and text messaging among other enhancements that are moving from today's society and culture into education (Goodale, 2007; Miller, 2005). Data acquisition, student interaction, and communication are at the center of all scientific activity and are key elements in science teaching and learning. The challenge for science teachers is the development of appropriate use strategies that enhance learning of the intended curriculum. A recent report by the Education Development Center questions current classroom practice in three areas: (1) the extensive time it takes in finding, developing skills and knowledge about, and using the best resources; (2) the high expense of materials and equipment; and (3) the difficulty of adapting technology resources to classroom and student needs (Hanson & Carlson, 2007).

TEACHERS' VIEWS OF LABORATORY WORK AND TECHNOLOGY IN TEACHING SCIENCE

Previous research has documented "typical" laboratory experiences and sequences as being about as effective in student science learning outcomes as are direct instruction and other typical noninquiry types of science pedagogy. Studies of "laboratory experiences," defined by the NRC as nontraditional, have been shown to be more effective for students in content mastery, scientific reasoning, and affective goal areas of science (Hofstein & Lunetta, 2004; Miller, 2004; Singer, Hilton, & Schweingruber, 2005). In a large experimental study, Freedman (1997) found that regular and continuous laboratory instruction significantly impacted students' attitude toward science and their science knowledge achievement. Laboratory instruction had significant effects on diverse students including those living in large urban settings and with limited English proficiency. Similar conclusions have been expressed in many recent reports of laboratory instruction (McComas, 2005; NRC, 2005).

Procedure for Teachers' Views

To better understand the importance of the laboratory and technology in the teaching of science a survey was conducted with practicing teachers and university faculty in the Southeast United States in spring 2007. Key questions considered were, "What are the perceptions of teachers of science toward the use and effectiveness of laboratory work and technology in the teaching of science?" and "What are teachers' patterns of use of laboratory work and technology in teaching?" The survey was designed to assess the awareness of, attitudes toward, and personal views of teachers regarding the extent to which laboratory work and the use of technology play a part in the teaching of science. The survey was e-mailed to a sample of classroom teachers of science, elementary grades through university level. The survey was sent to 105 teachers and faculty. Thirty-five surveys were sent to each of three groups: school systems, elementary and secondary teachers, and university faculty teaching entry level undergraduate courses. After reminders, the return rate was 88%, or 96 teachers and faculty. The return sample surveys closely matched the population with a small difference bias to the lower grades, where fewer responses were returned. The returned sample was judged appropriate to represent the population surveyed.

The survey consisted of one open response essay. The essay asked respondents to "provide a statement about your belief and actual use of the science laboratory (or technology) in your science teaching? (negative, neutral, or positive) The statement can be one sentence or a short paragraph of up to 150 words." Category themes and frequencies were determined for each questionnaire responses.

Results of Teachers' Views

The categories representing a majority of participants resulting from the analysis were motivation, concrete experiences, meaningful learning, and building procedural knowledge. No systematic differences in responses to beliefs and use were noted between the elementary, secondary, and university groups. Most teachers and faculty saw the use of the laboratory and technology as increasing *motivation*, making science real, and increasing students' interest in science lessons. One teacher explained "It is my belief that without lab work, science is dreary and surreal to students; in contrast, with an active lab component, science is real, relevant, fun, and meaningful" (VE, secondary science teacher, Tuscaloosa, Alabama).

There was a great amount of discussion of the remaining categories. Mentioned consistently throughout were statements indicating teacher and faculty understanding that use of the laboratory and technology makes science *clearer and is a way to connect the practice of science to theory* and of science making connections between the concrete and more abstract concepts of science. One teacher said, "I felt these experiments allowed them to really be able to relate with real life items in understanding the concept" (Fifth grade teacher #2, Shelby County, Alabama). Another teacher reported "I would never presume to teach science without lab activities and experiments. It would be like asking a student to describe riding a bike without ever letting them ride one." (Sabrina Stanley, secondary science teacher, Birmingham Alabama). Still another stated, "I felt these experiments allowed them to really be able to relate with real life items in understanding the concept" (Fifth grade teacher #3, Shelby County, Alabama).

Laboratory work and technology experiences were seen as necessary and helpful for *increasing meaningful understanding of the key concepts of science* and with connecting science to students' everyday lives.

> The authentic experiences gained in hands-on science labs cannot be duplicated in any other way. No video, lecture, or book in the world can come close to the depth of understanding gained by performing an experiment where students are able to gather their own evidence. (VE, secondary science teacher, Tuscaloosa, Alabama)

Teachers generally felt that without using laboratories or technology, the students would have missed in-depth learning of key science ideas. These experiences were considered crucial for student long term retention. Elementary teachers saw the need for the discourse involved in laboratory work.

They are excited about what's taking place, and they want to learn. I think there's a lot of activity going on. Lots of hands-on activity. Cooperative interaction between the students because they need to talk about what they are learning and discuss that with the others around and that way they are going to internalize it more and more as they talk about it. (Fifth grade teacher, Shelby County, Alabama)

At the elementary school level, teachers felt that reading or lecturing cannot be used alone. One fifth grade teacher stated "I know they need a lot of hands on. They need to see it. Lecture is not always enough" (Fifth grade teacher #1, Shelby County, Alabama). Another fifth grade teacher from Shelby County, Alabama noted "These experiments allowed them to really be able to relate with real life items in understanding the concept." At the secondary school level, integration of reading and lecture with hands-on investigations was held to be important.

When I first started teaching, I did not fully understand the importance of including lab in with lecture. As a new teacher, I was teaching under the misconception that terminology was the key to learning science and labs were just for fun. I could not have been more wrong. My students not only enjoyed the hands-on chemical experiments because they were entertaining or a break from the norm of lecture, but they truly began to understand science in the own personal lives through our labs. They were able to change their own misconceptions regarding science as they dissected starfish and frogs or observed chemical reactions with their own eyes. Without the inclusion of labs, my students would have missed the key to learning meaningful science. (SD, secondary science teacher, Charlotte, North Carolina)

The practice and development of *procedural knowledge* (low and high level skills) was viewed as an important outcome for using laboratory and technology activities in teaching and learning science. The experiences offer kinesthetic learning opportunities in what otherwise be a passive experience. The students' actions develop connections between the ideas of science and the actual processes involved in using science in their everyday lives. As one teacher stated, "The science lab encourages communication in the form of dialogue, writing and reading.... They also have the opportunity to improve their cooperative learning skills. Cooperation is a vital ability that students need. Students have the chance to learn from each other" (JH, fourth grade teacher, Cherokee County, Alabama).

Specific outcome areas of procedural knowledge included in the teachers' statements were communication of various types including speaking, writing, and reading; the science process skills of observing, classifying, identifying variables, and experimenting as well as others; complex inquiry skills such as problem solving, critical thinking, and creativity; and cooperative learning skills. Laboratory work, however, needs to be more than manipulating equipment. Many classroom science labs focus on mechanical procedures, such as biological slide preparation and chemistry pH measurement, to the detriment of relating the skill work to scientific principles or learning science concepts, The primary focus of laboratory work must be on ideas. These ideas were expressed by a secondary science teacher as:

> I think they need to have a voice in what they are doing and have an opportunity to have power and investigate some things. I think it's important for them to understand the scientific process and use it over and over again. I think building process skills is very important. It promotes critical thinking and it promotes knowledge of real life situations. I think it should include hands-on and inquiry-based things.... The more inquiry-based they do, the more they will start to develop questions about things and it will spark an interest in them and when they start questioning they will find themselves saying why is that ... why is that? It promotes critical thinking. (Middle School Teacher, Shelby County, Alabama)

Different student groups were seen to benefit especially from laboratory and technology experiences. The concreteness and the interactivity of laboratory experiences and the use of technology offer *diverse groups* of students different benefits. For gifted students, teachers noted that the use of technology and of the laboratory offered opportunities for modeling of abstract concepts. For lower performing students, these experiences offered concrete involvement with real objects to bridge the gap between the present knowledge of the student and the objective of the science lesson. One science teacher said it this way:

> When my special education students are given the time to use lab materials to make sense of the information we are discussing class, they seem to have a better understanding of the science content and retain that information for longer periods of time. (CP, secondary science teacher, Cullman County, Alabama)

The surveys received offered only a few *negative comments* on the use of the laboratory and technology in teaching science. Negative comments were included in about 25% of the total responses received. These statements centered on several key issues. One issue related to the unavailability of a specific space in which to perform laboratory or technology activities. This issue also concerned the inadequacy or unavailability of equipment and the persistent lack of software and consumable materials needed for replacements. Teachers were able to complete laboratories only by creative use of chairs and desks in regular classrooms and use of common, everyday materials to substitute for those typically used for representing science concepts.

A second issue involved students' lack of experience with effective laboratory learning. Typical student experiences in the laboratory, whether at the elementary or secondary levels, usually were disconnected from the science concept objectives of the curriculum. Or, the experiences were used to demonstrate concepts first introduced and then "proven" in the laboratory. When science teachers attempted to use inquiry laboratory activities involving investigation of problems leading to the understanding of a key science concept, the students treated the activities as fun but unimportant learning that should be assessed. Students' approach to laboratory work in the middle school was described by one teacher as

> We do not have a science lab at [the school] for student use. The only room that is designed to be used as a lab is being used as a classroom. When we do labs, I turn four desks together to make a square … I had great difficulty at the beginning of the year because there were no materials to use such as beakers, burners, microscopes, slides etc. I was able to purchase dissecting materials myself. The kid bought pigs to dissect the first couple of years. I had to use my instructional funds to purchase the needed materials…. We didn't have textbooks either…. We had two functional microscopes which limited the amount of use per science teacher and within the class. (JJ, secondary science teacher Tuscaloosa, Alabama)

A third issue concerned the need for professional development to assist teachers in establishing effective learning in the laboratory and in making connections between the laboratory and other activities such as lecture and readings involved with learning the same concept. The kind of teaching evident in many school laboratories neither promotes students' learning of science concepts, nor does it help students to learn about the practices of science (Berry, Gunstone, Loughran, & Mulhal, 2001). Changes are needed in the purpose for planning and implementing science laboratory work for students. A secondary science teacher described the students' experience with laboratory experience in this way:

> I personally felt that the students participated more during lab time and that they were more interested in the lab activities. I did notice that no matter how hard I tried to do a lab that introduced and/or reinforced the material, most students failed to see any relationships between the labs and the material. All they were concerned about was doing what was necessary to get a grade for the lab. It's as if they view the labs as a free day as opposed to a learning day. I assumed that this was because I was teaching lower level students until I substituted for an advanced chemistry teacher one day. I

overheard the students talking about how the science teacher thought they understood the labs, but they actually never knew how they related to what they were covering in class. (Secondary science teacher, Tuscaloosa, Alabama)

A fourth issue centered on the importance of laboratory use in the science curriculum and in the organization of the school schedule. Since all teachers and subjects are treated the same way in secondary schools, it is difficult to teach science classes differently from the way English classes are taught. The expectation was to use the same pedagogy, level of student action in the classroom, and time schedule. A common statement made by the teachers follows this pattern:

The utilization of the science laboratory is the most beneficial aspect of science exploration. It is to science teachers, the best way for students to connect practice to theory. However, because of multiple overloads, new mandates, and time constraints, science teachers find it extremely difficult to attend the lab as much as they need. (Secondary science teacher, Huntsville, Alabama)

Conclusion on Teachers' Views

Responses to the two questions guiding the survey, in summary, are (1) the perceptions of teachers of science toward the use and effectiveness of laboratory work and technology in the teaching of science are similar to those found in policy documents of the past 10 years, and (2) The science laboratory and technology are seen as an integral part of science teaching. The teachers' patterns of use of laboratory and technology experiences in teaching were centered on four areas: motivation, concrete experiences, meaningful learning, and building procedural knowledge. What was missing from the surveys were statements and terminology that affirmed the importance of inquiry learning and use of investigation to introduce and apply the science concepts in the curriculum. The language used throughout the surveys returned suggested that teachers view laboratory and technology as only assisting them in explaining the science concepts introduced in the textbooks and lectures. The laboratory and technology experiences were not envisioned as a primary way to introduce and teach science concepts. The statements mentioned hands-on learning with concrete concepts but it was rare to find statements that outlined an inquiry laboratory strategy being accomplished where students are constructing their own meaning taking control of their own learning.

Laboratory Work and Using Technology in Science Teaching and Learning in Today's Schools and Universities

A fundamental goal in science teaching is to help students become engaged and self-directed in their learning (NRC, 1996). This is a change from the past and is the primary factor in planning for inquiry learning. In addition, knowledge in science and technology is changing and expanding rapidly as new knowledge is acquired. Science teachers must be able to relate and accommodate to this change by keeping up their own knowledge and skills and using the new knowledge and skill to guide instructional practice in their own teaching.

A range of laboratory methods are used in teaching science that involve differing teacher and student roles (see Figure 1.1). Different laboratory methods vary importantly in the amount of student control

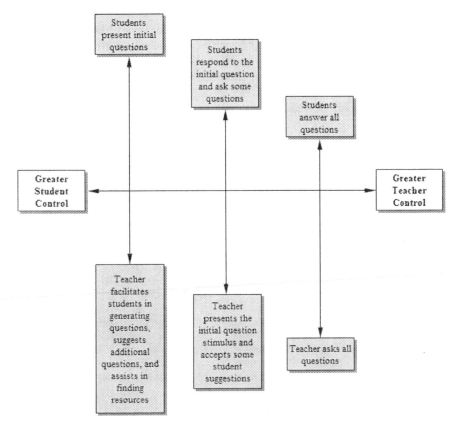

Figure 1.1.   Role taken in laboratory activities: A continuum.

during the learning process. Laboratory activities may involve massing liquids, reading laboratory procedures in the lab exercise book, watching a demonstration, or following instructions of the teacher written on the whiteboard in a step by step process. Contrasting activities are creating a journal about an experiment, designing an investigation based on an observed problem, completing a science fair project, or asking a question and being encouraged to follow up the answer through one's own investigations. The difference in these two sets of activities involves the amount of control the student has in performing them.

The amount of control students have over their science learning activities is an important factor in helping them create meaningful learning (Simsek, 1993). Laboratory methods for teaching science can be grouped into four categories based on amount of student control. Sequenced from least to greatest student control, the categories are: confirmation, structured inquiry, guided inquiry, and open inquiry (see Table 1.1). The first falls under *direct instruction* method while the last three are forms of *inquiry laboratory* methods (see Table 1.1). Each laboratory method is particularly effective and appropriate in helping students accomplish a "specific" goal of science learning (see Figure 1.2). If used exclusively, inquiry laboratory methods are more effective than others for accomplishing higher levels of learning outcomes. Higher levels of inquiry in the laboratory result in higher cognitive skills, procedural knowledge (skills), and affective level learning outcomes for students.

Table 1.1.   Student and Teacher Roles During Laboratory Activities

Laboratory Activity Type	Role of Student	Role of Teacher
Confirmation	Attends to what the teacher is doing and saying, performs activity as instructed for result already known	Provides question, procedure, materials, expected results, and determines appropriateness of answer
Structured inquiry	Investigates, constructs answers, and determines appropriateness to the question	Provides question, procedure and materials to carry out an investigation and facilitates students in their role.
Guided inquiry	Plans investigation, performs investigation, constructs answers, and determines appropriateness to the question	Provides question and materials to carry out an investigation and facilitates students in their role
Open inquiry	Determines question, plans investigation, performs investigation, constructs answers, and determines appropriateness to the question	Facilitates students in their role

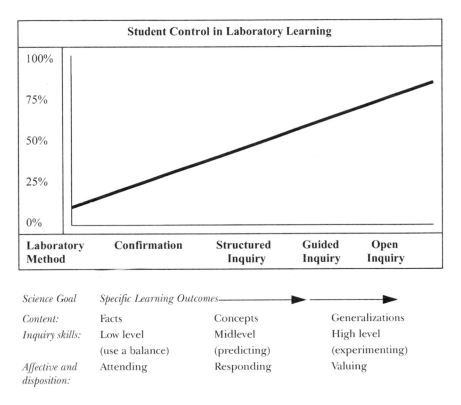

Student Control in Laboratory Learning				
Laboratory Method	**Confirmation**	**Structured Inquiry**	**Guided Inquiry**	**Open Inquiry**

Science Goal   *Specific Learning Outcomes* ⟶ ⟶

Content:	Facts	Concepts	Generalizations
Inquiry skills:	Low level	Midlevel	High level
	(use a balance)	(predicting)	(experimenting)
Affective and disposition:	Attending	Responding	Valuing

Figure 1.2.   Matching student laboratory involvement and learning outcome.

It is important to understand how and when each laboratory method is to be used. No science laboratory method is bad or good by itself. However, the use of confirmation methods for a large proportion of any science laboratory unit is inappropriate (Collins, 1997). Confirmation methods also are inappropriate for teaching science concepts, generalizations, or higher order thinking skills meaningfully. However, confirmation is appropriate for the times when teachers want students to learn how to use a digital balance, measure volume in a graduated cylinder, or perform the procedure for staining an animal cell for observation under a microscope. Selecting the laboratory procedure as a match for the learning outcome intended is part of the expert teacher's laboratory and technology planning process.

The integration of laboratory and technology experiences with the everyday teaching of key science curriculum concepts is a critical condition for laboratory learning. For example, laboratory and/or technology experiences should be scheduled before key concepts are introduced in a lesson. This introduction is to be done as a problem to be investigated and to help students relate their own prior knowledge to the idea. Next, in the lesson, laboratory and technology experiences can be scheduled at the same time the key ideas are explained. Still later, the laboratory and technology experiences can be scheduled near the end of the lesson to allow students to apply the new ideas to gain a deeper understanding of the meaning of the concept. The laboratory and the use of technology also could provide a vehicle for facilitating students' transfer of the explained concepts to a new setting and adaptation of the concept for a different use in a new context. This strategy of integration brings together both theory and research-based practice. The strategy represents the planning sequence used in learning cycle lesson design (Sunal, 2004; Sunal & Sunal, 2003).

Analysis of Inquiry Potential of Laboratory Materials

Unfortunately, science laboratory materials and exercises usually provided to teachers of science, K-16, are still centered on traditional methods of the past decades. What is needed is a systematic way of critiquing these materials and adapting them to more effective, research-based, laboratory methods.

Several authors have suggested a process that assists teachers in adapting student laboratory work and technology experiences from traditional cookbook or confirmation activities into inquiry-orientated investigations (Herron, 1971; Knecht, 1974; McComas, 2005; NRC, 2000a; Schwab, 1962; Tafoya, Sunal, & Knecht, 1980; Volkmann & Abell, 2003). The adaptation process has several steps: (1) analysis of laboratory and/or technology learning materials and student work following the essential features of inquiry learning (NRC, 2000a), (2) identification of elements needing modification, and (3) adapting the materials and creating appropriate experiences engaging students in inquiry related to key science concepts of the lesson.

Analysis of Laboratory and/or Technology Learning Materials and Student Work

In keeping with policy statements on the goals of science laboratory work and use of technology, inquiry for this adaptation process was operationally defined as a strategy for teaching that:

1. is devoid of authoritarian answers to science questions
2. provides empirical verification by students of knowledge claims (uses research-based evidence)

3. involves active student investigation with diverse materials in a variety of settings

4. includes student participation in all phases of knowledge generation at their [developmental] level

5. involves the complete inquiry process—assuming, observing, inferring, hypothesizing, testing, and revising ideas and concepts on the basis of new information (Knecht, 1974; Tafoya, Sunal, & Knecht, 1980).

Analysis follows the NRC's (2000a) description of the essential features of student inquiry in science teaching. These features involve students in engaging questions, searching for and use of evidence, developing explanations from evidence, connecting and determining validity of explanations, and dissemination of findings (NRC, 2000a, p. 25) (see Table 1.2).

Each of the essential elements can vary in the amount of student control over learning that is allowed in the lesson. Schwab (1962) and Herron (1971) describe this control variation as an appraisal of how much information the student has when the activity begins. Level 0 involves providing students with all of the information needed to answer the question asked for the laboratory. Level 2 provides some of the information. Level 3 asks students to provide and perform all aspects of the laboratory exercise. Connection to the NRC's essential elements of inquiry (see Table 1.3) is best described by Knecht (1974) who identified types of activities reflecting different degrees of autonomous inquiry by students as confirmation, structured inquiry, guided inquiry, and open inquiry.

It is possible to use these basic definitions to construct a rubric outlining various aspects of the laboratory materials to complete the analysis. After the teacher has secured all laboratory materials including laboratory lesson plans, student handout sheets, and other materials related to implementing the lab, analysis can begin. The two part rubric "Analysis of Inquiry Potential Rubric for Determining the Level of Inquiry

Table 1.2. Essential Features of Inquiry

Students are Involved in Inquiry if the Following Behaviors and Observed
1. Learner engages in scientifically oriented questions
2. Learner gives priority to evidence in responding to questions
3. Learner formulates explanations from evidence
4. Learner connects explanations to scientific knowledge
5. Learner communicates and justifies explanations

Table 1.3.   Types of Activities Reflect Different Degrees of Autonomous Inquiry by Students

Herrin Taxonomy Level (1971)	Inquiry Lab Method Category (Tafoya, Sunal, & Knecht, 1980)
0	(a) *Confirmation*. A concept or principle is presented and the student performs some exercise to confirm it. She/he knows what is supposed to happen and the procedure has been carefully outlined for the student to follow.
1	(b) *Structured inquiry*. The student is presented with a problem by the teacher but does not know the results beforehand. Procedures are outlined. Selection of activities and materials is structured to enable the student to discover relationships and to generalize from data collected.
2	(c) *Guided inquiry*. The teacher only provides the question or problem enabling the student determine the procedure and conduct activities to discover relationships and to generalize from data collected. The teacher facilitates the rest of the students' decisions and activities.
3	(d) *Open inquiry*. In this level of inquiry the question or problem to investigate is determined the student. The student directs his/her own procedures and methods of collecting data from which concepts or principles are discovered and generalized and conclusions are drawn. The teacher facilitates the students' decisions and activities (Tafoya, Sunal, & Knecht, 1980).

in Laboratory Materials and Student Work (AIP)" provides for a sequence of decisions to be made regarding (1) the key laboratory question or problem that guides the investigation and (2) analysis of laboratory methods to be used (see Table 1.3). The basic question asked and responded to in part one is "Is the scientific, or is it an assertion that cannot be verified through inquiry investigation activities?" Four questions need to be answered in this section for the analysis. The basic question asked and responded to in part two is "Do the laboratory materials and the instructional procedure engage the students in an inquiry investigation?" Five questions need to be answered in this section for the analysis.

Identification of Elements Needing Modification

When completing the AIP rubric the teacher documents the elements of the laboratory lesson that need modification. Revising the key question or problem should come first (see AIP rubric item 4). When the question is revised and complete, revision of the type of laboratory method should be completed next (see AIP rubric item 8). Finally, modification of the complete teacher lesson plan and the student handouts comes last (see Table 1.4).

Adapting the Materials and Creating Appropriate Experience.

Once the elements needing modification have been identified in the AIP analysis, adaptation of the laboratory materials can be completed by creating appropriate experiences that engage students in inquiry related to the key science concepts of the lesson. While making these adaptations having a copy of the *Essential Features of Inquiry* (see Table 1.2) is useful. These features will help guide the questions to be asked about the appropriateness of the laboratory materials.

Example Analysis and Modification of a Laboratory Lesson Using AIP

The laboratory lesson scenario found in Table 1.5 is a sample lesson that has been analyzed as being in need of adaptation from a traditional laboratory to one that engages the students in more inquiry type laboratory activities. The reader may want to try to complete their own analysis before reading further. A second scenario (see Table 1.6) provides an adapted laboratory lesson sample that followed the analysis results as a guide on making changes. The major change was made by revising the laboratory question and placing it at the beginning of the laboratory activity. Other changes made were providing additional time for inquiry to take place, focusing lab questions on statements that led to inquiry investigations, introducing explanations and terminology after the student experiences with the phenomena, giving students greater control over their learning activities, and structuring the lesson throughout to be based on the use evidence to create meaning and explanations. As a general rule, changes should be made reflecting the guiding essential inquiry features questions in item #5 of the AIP (see Table 1.4).

Changes can be made to adapt laboratory lessons reflecting the recent emphases in developing a higher level of importance for laboratory work described in the science education literature. These sources have provided guidelines that encourage (1) engagement in long-term investigations, (2) incorporation of a more authentic nature of science foundation of activities, (3) development of investigations outside of the classrooms on the school grounds and during field trips, (4) use of laboratory work to begin science lessons where concepts are introduced and students prior knowledge is connected, (5) provision of opportunities for students to make their own decisions during laboratory work, (6) assessment of laboratory work using performance assessment rather than paper and pencil testing, (7) use of cooperative groups of students in conducting investigations, and (8) assessment of group work in addition to individual work

Table 1.4.   Analysis Scheme for Determining Level of Inquiry in Laboratory Activities

Laboratory Question or Problem Analysis:

Is the scientific statement used to begin the laboratory activity an inquiry (experimental) question or problem or is it an assertion that cannot be verified through inquiry investigation activities?

Use the next four questions to determine the nature of the scientific statement used by the teacher to begin the laboratory activity.

1. Is the scientific statement theoretical? (i.e., Does the sentence use terms which the students can't understand? Or does it tell them something which they can't verify?)

    a.   If yes, the statement must be modified to raise the inquiry level of the laboratory inquiry. Then go to 5.

    b.   If no, go to 2.

2. Is the scientific statement methodological? (i.e., Does it state something about the nature of science or the scientific method?)

    a.   If yes, the statement must be modified to raise the inquiry level of the laboratory inquiry. Then go to 5.

    b.   If no, go to 3.

3. Is the scientific statement a definition?

    a.   If yes, the statement must be modified to raise the inquiry level of the laboratory inquiry. Then go to 5.

    b.   If no, go to 4.

4. Is the scientific statement based on inquiry (experimental)? (i.e., Does it state something which the student can understand and verify?)

    a.   If yes, go to 5.

    b.   If no, modify the statement into a question that can be appropriately tested through inquiry investigation activities and go to 5.

Analysis of Laboratory Methods:

Do the laboratory materials and the instructional procedure engage the students in an inquiry investigation?

5. Do the laboratory materials describe conditions (NRC, 2000) which will allow the students (on their own) to investigate or design a test and make a response for the question or problem in the following ways?

    •   Is the learner expected to engage in scientifically oriented questions? Can the learner ask or change the question to be investigated to make them more relevant?

    •   Is the learner expected to give priority to evidence in responding to questions? Is the learner given opportunities to design and to collect primary or secondary data related to scientifically oriented questions?

    •   Does the learner formulate explanations from evidence observed during investigation?

- Is the learner expected to construct and connect these explanations to scientific knowledge? Is the learner expected to explain the reasoning behind their explanations?
- Is the learner expected to compare, revise communicate and justify their explanations in the laboratory lesson?
   a. If yes is given to all questions a) to e), go to 7.
   b. If no, go to 6.

6. Can the teacher and students (together with the group or whole class) design an investigation (experiment) to answer the scientific question or problem using the 5 conditions (NRC, 2000) specified in item 5?

   a. If yes, tally and go to 7.

   b. If no, reclassify this scientific question or problem by going to item 1 and either modify or eliminate.

7. How will the student answer the question or problem?

   a. Using a laboratory investigation activity? If yes, go to 8.

   b. Using an examination or analysis of photographs, illustrations, digital media, secondary data, etc.? If yes, and if appropriate, go to 9.

   c. By teacher demonstration? If yes, and if appropriate, go to 9.

   d. By a teacher lecture or another authority source? If yes, reclassify this scientific question or problem by going to item 1 and either modify or eliminate.

   e. By some other procedure? If yes, indicate method and if appropriate go to 9.

8. What type of laboratory method will be used? (see Table 1.1)

   a. Confirmation? If yes, reclassify this scientific question or problem by going to item 1 and either modify or eliminate.

   b. Structured inquiry? If yes, and if appropriate, go to 9.

   c. Guided inquiry? If yes, and if appropriate, go to 9.

   d. Open inquiry? If yes, and if appropriate, go to 9.

   e. By some other procedure? If yes, indicate method and if appropriate go to 9.

9. What procedures will be used to interpret or evaluate the results of the inquiry? Make revisions to the laboratory lesson plan and student materials that are appropriate to the analysis results.

   a. Individual data collection?

   b. Individual laboratory report?

   c. Pooling of laboratory data collection and reporting for group or class examination?

   d. Cooperative group laboratory report?

   e. Some other procedure?

Table 1.5.   Sample High School Biology Lesson #1

This high school biology lesson was in the middle of a unit on cells. In the previous lesson, students had read in their textbooks about the process of diffusion and the results of conducting an experiment in which one placed either starch or sugar solution in dialysis tubing and then submerged the tubing in a beaker of water with indicator. It was explained that molecule size and transport across cell membranes would be determined by the large size of the starch molecule. The purpose of today's membrane lab lesson was to begin show how the large starch molecule affects the transport across cell membranes as discussed in their textbook.

Following a review lecture on the reading, the teacher supplied each group of students with a copy of the procedures for conducting the membrane lab and a related worksheet, both from their laboratory workbook. The worksheet required students to complete the lab for both starch and sugar and to describe their results at the end. In addition, there was an additional exercise with a matching section related to vocabulary from the textbook.

Most students had just begun the worksheet when the teacher asked who was confused about the worksheet. When many students raised their hands, the teacher led a whole class in a tutorial about how the lab was to be carried out. The students then returned to the lab and the worksheets for a few more minutes. During the lab the teacher circulated to ask questions and offer assistance. The teacher asked questions that engaged the students, and probed further at times. Once students had finished the laboratory and the worksheet they discussed it as a class, with students volunteering their answers and the teacher writing them on an overhead. Finally, the teacher summarized the results and made the connection to the explanation in their textbook.

(McComas, 2005). Many practical examples of integrating technology in inquiry laboratory work in science teaching and learning can be found in journals such as the *Science Teacher* (e.g., Hisim, 2005) and on the Internet.

The AIP instrument provides an objective procedure for analyzing laboratory materials on the basis of their potential contribution to well recognized goals of science education. It is, therefore, a highly significant tool for use in the selection and adaptation of laboratory and in general laboratory curricular materials. Such analysis can remove the decision-making process from the realm of vague intuitions to one where materials may be selected and or modified with a high degree of predictive validity for achieving the chosen goals of science education with students.

SUMMARY

This chapter has explored the importance of the use of the science laboratory and technology for teachers and students in today's science classroom. The focus for this exploration was the historical evolution, viewpoints of policy documents, and expectations of stakeholders

Table 1.6.   Sample High School Biology Lesson #2

This high school biology lesson was in the middle of a unit on cells. In the previous laboratory lesson, students, in cooperative groups, had conducted a membrane lab in which they placed either starch or sugar solution in dialysis tubing and then submerged the tubing in a beaker of water with an indicator. The question guiding students in this lab was, What variables affect the passage of dissolved liquids through a cell wall? Predictions were made first. Observation and recordings were then completed on the events. A discussion of the predictions and observed results were made within small groups as the class session ended.

The next lesson began with a discussion of the predictions and observed results between the lab groups in a whole class setting. Hypotheses were suggested and written on the board. The purpose of today's lesson was to begin to draw together student ideas, especially about the variables molecule size and transport across cell membranes. The teacher asked the students, in their lab groups, to predict what they expected to have happened with their lab based on the molecule size (i.e., whether the starch and/or sugar would have diffused across the membrane). The teacher then asked, based on their particle size prediction, to explain why it works that way.

Through discussion and later small group reports, the groups examined their data and discussed whether their predictions were right or wrong. The teacher then led the entire class in a discussion about what had happened in the experiment. Students suggested additional hypotheses, and the class discussed methods for testing them. As needed, the teacher chimed in with suggestions (e.g., using test tape to measure sugar content), but his role was primarily providing lab techniques that would enable the students to test their ideas and prodding the groups to make sure they conducted enough tests to fully explain what had happened.

During this segment of the lesson the students were in charge of their investigations and doing the majority of the intellectual work. The teacher kept to her role of facilitator, questioning students and giving them some suggestions for lab tests. The teacher guided the students as they finished making observations and analyzing the data, asking questions that pushed students to examine their results and to provide evidence for their conclusions. Examples of questions asked by the teacher are: "How could we test if there is still sugar in the reservoir?" "Why didn't it (the iodine indicator) reach an equilibrium?" and "How do you know?" The teacher also introduced new vocabulary (events, materials, objects, processes etc.) to the class after students experienced the events and were discussing the results. For example, as the students were trying to explain what had happened to the sugar in their experiment, the teacher interjected to the whole class "I hear you discussing, let me introduce a term: equilibrium." The teacher was able to introduce new content into the discussion in the context of the investigation.

After the groups had finished all of their tests, the teacher gave them an assignment to write a story about a paramecium that lived in the local freshwater river who decided to go see his girlfriend who lived in the ocean. The groups were instructed to write about his trip and what he would experience. The teacher supplied them with a list of eight vocabulary words related to transport across a membrane that they had to use in the story. The groups were told that the teacher would call on one group member to read and explain their story to the class the next day, so they all needed to understand the concepts they included. The students spent the remainder of the class period working on their stories. This activity provided time for the students to bring together what they knew about transport across a membrane and supply it to organisms living in their local river.

*Modified from a science classroom scenario originally written by a classroom observer in a national study, "Inside the Classroom," conducted by Horizon Research Inc. (Weiss, Pasley, Smith, Banilower, & Heck, 2003, p. H 29).

involved in planning and implementing science laboratory and technology experiences. A final result of this exploration was the description of a process for helping teachers gather objective data when making critical judgments on planning, use, and adaptation of existing laboratory materials based on the goals and recognized principles of good science education. The Analysis of Inquiry Potential rubric instrument was illustrated, along with an example of its use, as a guide and aid to assist teachers and other decision makers in designing more effective materials and procedures for using laboratory work and technology experiences.

REFERENCES

American Association for the Advancement of Science. (1993). *Benchmarks for science literacy: Project 2061*. New York: Oxford University Press.

Berry, A., Gunstone, R., Loughran, J., & Mulhal, P. (2001). Using laboratory work for purposeful learning about the practice of science. In H. Behrendt (Ed.), *Research in science education- past, present, and future* (pp. 38-318). Norwell, MA: Kluwer.

Center for Education Policy (2007). *Choices, changes, and challenges: Curriculum and instruction in the NCLB era*. Retrieved July 24, 2007, from http://www.cep-dc.org/index.cfm?fuseaction=document .showDocumentByID&nodeID=1&DocumentID=212

Collins, A. (1997). National science education standards: Looking backward and forward. *Elementary School Journal, 97*(4), 299-313.

Freedman, M. (1997). Relationship among laboratory instruction, attitude toward science, and achievement in science knowledge. *Journal of Research in Science Teaching, 34*(4), 343-357.

Goals 2000: Educate America Act, 20 U.S.C. & 5801 et.seq. (1994)

Goodale, G. (2007). Incase of emergency, play video game. *Christian Science Monitor*. June 6, 2005, Retrieved April 16, 2007, from http://www .csmonitor.com/2005/0606/p11s01-legn.html

Hanson, K., & Carlson, B. (2007). *Effective access: Teachers use of digital resources in STEM teaching*. Boston: Education Development Center. Retrieved April 16, 2007, from http://www2.edc.org/GDI/publications _SR/EffectiveAccessReport.pdf

Herron, M. (1971). The nature of scientific inquiry. *School Review, 79*, 171-212.

Hisim, N. (2005). Technology in the lab: Part 2—Practical suggestions for using Probeware in the science classroom. *Science Teacher, 72*(7), 38-41.

Hofstein, A., & Lunetta V. (2004). The laboratory in science education: Foundations for the twenty-first century. *Science Education, 88*, 28-54.

Hudson, S., McMahon, K., & Overstreet, C. (2000). The national survey of science and mathematics education. Chapel Hill, NC: Horizon Research.

Knecht, P. (1974). *A model to facilitate the assessment of epistemological quality in elementary science programs*. Unpublished doctoral dissertation, Michigan State University.

Martens, M. L. (1992). Inhibitors to implementing a problem solving approach to teaching elementary science: A case study of a teacher in change. *School Science and Mathematics, 92*(3), 153.

McComas, W. (2005). Laboratory instruction in the service of science teaching and learning. *Science Teacher, 72*(7), 24-29.

Miller, M. (2005). Technology in the lab: Part 1—What research says about using Probeware in the science classroom. *Science Teacher, 72*(7), 34-37.

Miller, R. (2004). *The role of practical work in the teaching and learning of science. Paper prepared for the Committee on High School Science Laboratories: Role and Vision.* Retrieved September 15, 2005, http://www7.nationalacademies.org/bose/June3-4_2004_High_School_Labs_Meeting_Agenda.html

National Commission on Excellence in Education. (1983). *A nation at risk: The imperative for educational reform.* Retrieved April 16, 2007, from http://www.ed.gov/pubs/NatAtRisk/index.html

National Education Association. (1894). *Report of the Committee of Ten on secondary school studies.* New York: American Book Company.

National Research Council. (1996). *National Science Education Standards.* Washington, DC: National Academy Press.

National Research Council. (2000a). *Inquiry and the National Science Education Standards.* Washington, DC: National Academy Press.

National Research Council. (2000b). *How people learn: Brain, mind, experience, and school* (Expanded edition). Washington, DC: The National Academies Press.

National Research Council. (2005). *America's lab report: Investigations in high school science.* Washington, DC: National Academy Press.

National Science Teacher's Association. (1990). *NSTA position statement. Science/technology/society: A new effort for providing appropriate science for all.* Retrieved April 16, 2007, from http://www.nsta.org/about/positions.aspx

National Science Teacher's Association. (2007). *Integral role of laboratory investigations in science instruction.* Retrieved April 16, 2007, from http://www.nsta.org/about/positions.aspx

No Child Left Behind Act of 2001, 20 U.S.C. & 6301 et seq (2002).

Psillos, D., & Niedderer, H. (2002). Issues and questions regarding the effectiveness of labwork. In D. Psillos & H. Niedderer (Eds.), *Teaching and learning in the science laboratory* (pp. 21-30). Boston: Kluwer.

Rudolph, J. (2002). *Scientists in the classroom: The cold war reconstruction of American science education.* New York: Palgrave.

Rutherford, F., & Ahlgren, A. (1989). *Science for all Americans.* New York: Oxford University Press.

Schwab, J. (1962). The teaching of science as inquiry. In P. F. Brandwine (Ed.), *The teaching of science* (pp. 1-103). Cambridge, MA: Harvard University Press.

Simsek, A. (1993). The effects of learner control and group composition in computer-based cooperative learning. *Proceedings of selected research and development presentations at the annual convention of the Association for Educational Communications and Technology sponsored by the Research and Theory Division.* New Orleans, Louisiana: Association for Educational Communications and Technology, 1-39. ERIC Document Reproduction Service No. ED 362 205

Singer, S., Hilton, M., & Schweingruber, H. (Eds.). (2005). *America's lab report: Investigations in high school science. Committee on High School Science Laboratories: Role and Vision, Board on Science Education, Center for Education, Division of Behavioral and Social Sciences and Education, National Research Council.* Washington, DC: The National Academies Press.

Sunal, D. (2004). Innovative pedagogy for meaningful learning in undergraduate science. In D. Sunal & E. Wright (Eds.), *Research in science education: Reform in undergraduate science teaching for the 21st century* (pp. 85-122). Greenwich, CT: Information Age.

Sunal, D., & Sunal, C. (2003). *Teaching elementary and middle school science.* Columbus, OH: Merrill Prentice Hall.

Sunal, D., Wright, E., & Sunal, C. (2008). *Science in the secondary school.* Boston: Allyn and Bacon.

TaFoya, E., Sunal, D. W., & Knecht, P. (1980). Assessing inquiry potential: A tool for curriculum decision makers. *School Science and Mathematics, 80*(1), 43-48.

Volkmann, M., & Abell, S. (2003). Rethinking laboratories. *Science Teacher, 70*(6), 38-41.

Weiss, I., Pasley, J., Smith, P. S., Banilower, E., & Heck, D. (2003). *Inside the Classroom.* Chapel Hill, NC: Horizon Research.

PART I

DEVELOPMENT AND USE OF SCIENCE LABORATORY WORK IN LEARNING AND TEACHING

INTRODUCTION

The position statement of the National Science Teachers Association (NSTA) indicates the association considers the science laboratory central to science learning and teaching.

A hallmark of science is that it generates theories and laws that must be consistent with observations. Much of the evidence from these observations is collected during laboratory investigations. A school laboratory investigation (also referred to as a lab) is defined as an experience in the laboratory, classroom, or the field that provides students with opportunities to interact directly with natural phenomena or with data collected by others using tools, materials, data collection techniques, and models.

Throughout the process, students should have opportunities to design investigations, engage in scientific reasoning, manipulate equipment, record data, analyze results, and discuss their findings. These skills and knowledge, fostered by laboratory investigations, are an important part of inquiry-the process of asking questions and conducting experiments as a way to understand the natural world.

While reading about science, using computer simulations, and observing teacher demonstrations may be valuable, they are not a substitute for laboratory investigations by students.

For science to be taught properly and effectively, labs must be an integral part of the science curriculum. The National Science Teachers Association (NSTA) recommends that all pre K-16 teachers of science provide instruction with a priority on making observations and gathering evidence, much of which students experience in the lab or the field, to help students

The Impact of the Laboratory and Technology on Learning and Teaching Science K-16, pp. 31–33

develop a deep understanding of the science content, as well as an understanding of the nature of science, the attitudes of science, and the skills of
scientific reasoning.

Furthermore, NSTA is committed to ensuring that all students-including
students with academic, remedial, or physical needs; gifted and talented
students; and English language learners-have the opportunity to participate
in laboratory investigations in a safe environment. (NSTA, 2007, p. 1)

The first section of this volume addresses research in practical science
work for students and the barriers teachers typically face in implementing
quality practical experiences. In chapter 2, William Sandoval addressed
effective design and subsequent implementation of laboratory experiences in high schools. In the third chapter, Deborah Hanuscin provided
research on the use of science laboratory facilities in elementary classrooms. Both authors note that limited resources for laboratory instruction
and prevalence of traditional, "cook-book" laboratory activities as barriers
to student science learning.

Sandoval reviewed findings from the committee established by the
National Research Council (NRC) to review, analyze, and report on
laboratory activities in America's high schools. His review of the
committee's report, *America's Lab Report: Investigations in High School
Science (ALR)* (Singer, Hilton, & Schweingruber, 2005), along with other
related research concerning the justification for providing laboratory
experiences form a foundation for the recommendations made in the
chapter. The chapter opens with concerns of the paucity of credible
research in the effectiveness of laboratory experiences provided by
researchers like Hofstein and Lunetta in 1982 and later by Lazarowitz and
Tamir in 1994. Generally, the research base indicated students benefited
little from traditional teacher-centered laboratory activities. In contrast,
Sandoval reported that recent research over the past 20 years has
provided evidence that students tend to benefit more from inquiry-based
laboratory experiences. Following a brief overview of the research base,
Sandoval   advocated   for   the   implementation   of   the   NRC's
recommendations, described in the ALR report, for effective design of
laboratory activities: (1) communicate a clear purpose for the laboratory
experience, (2) carefully sequence instruction to maximize student
learning, (3) maintain an integrated focus on science concepts and
laboratory practice, and finally, (4) exemplary design of laboratory
experiences includes ongoing discussion and reflection on the experience
as an integral part of the learning process.

. The chapter concludes with case studies of effective laboratory
instruction using the guidelines of the NRC and, in addition, places an
emphasis on the importance of laboratory experiences in the
development of student understanding of the nature of science. Finally,

Sandoval reiterated the need for extensive, well-designed research on the effectiveness of laboratory experiences for high school students.

In chapter 3, Hanuscin outlined the utilization of separate facilities for laboratory experiences in elementary classrooms. The chapter begins with barriers to effective elementary laboratory instruction and ways in which separate laboratory facilities reduce these barriers. Hanuscin reported little priority was given to science, both in terms of instructional time and resources, and even when provided with separate laboratory facilities and adequate consumable resources, lack of teacher preparation and/or priority to science limits science instruction in the elementary classroom. Included in the chapter is a discussion of the positive impact of elementary science specialists, who are trained to teach science in the elementary classroom and typically place priority on science. The chapter concludes with a research agenda for effective laboratory teaching in the elementary classroom.

Part I begins with effective instruction in the high school science laboratories and concludes with its counterpart in elementary schools. Sandoval and Hanuscin reported similar barriers to effective laboratory instruction including traditional pedagogy and limited resources. In each chapter, the reader will find exemplars for laboratory experiences in both secondary and elementary classrooms.

REFERENCES

National Science Teacher's Association. (2007). *NSTA position statement: Integral role of laboratory investigations in science instruction.* Retrieved April 16, 2007, from http://www.nsta.org/about/positions.aspx

Singer, S., Hilton, M., & Schweingruber, H. (Eds.). (2005). *America's lab report: Investigations in high school science. Committee on High School science laboratories: Role and vision, Board on Science Education, Center for Education, Division of Behavioral and Social Sciences and Education, National Research.* Washington DC: The National Academies Press.

CHAPTER 2

DESIGN PRINCIPLES
FOR EFFECTIVE
LABORATORY INSTRUCTION

William A. Sandoval

Laboratories have been a fixture in science education for more than a century, yet their value has been debated almost that entire time. In 2005, the National Research Council (NRC) released *America's Lab Report*, synthesizing recent inquiry-oriented research efforts that use laboratory experiences more effectively than is typical. This chapter comprehensively describes four design principles for effective laboratories identified in the NRC report: clearly communicated purposes; careful sequencing within the flow of instruction; integrated focus on concepts and practice; and ongoing discussion and reflection. Each principle is described, given a philosophical rationale, clarified and justified with evidence from cognitive developmental research. The chapter concludes with examples of how these design principles have been used effectively. The aim is to help science educators and curriculum developers to apply current research to the development of quality laboratory experiences for children.

The Impact of the Laboratory and Technology on Learning and Teaching Science K-16, pp. 35–56
Copyright © 2008 by Information Age Publishing

INTRODUCTION

Laboratories have been a feature of science education for more than a century, yet their value has been debated many times over. What do students really get out of laboratory activities in their science classes? Are they necessary components of good science instruction? How can labs be improved? In 2004, the National Science Foundation asked the National Research Council to provide a definitive answer to these questions. The study committee on high school laboratories published its findings in 2005 as *America's Lab Report: Investigations in High School Science (ALR)* (NRC, 2005). The committee found that definitive answers to the first two questions posed above are hard to come by. Laboratory experiences are proposed to meet a number of goals of science education, but the existing research base often provides inconclusive evidence of the benefits of labs. At the same time, the committee suggested that science education "would not be about science if it did not include opportunities for students to learn about both the process and the content of science" (NRC, 2005, p. 3). While the report notes that labs are only one potential way to provide these opportunities, the committee found that a number of research-based efforts to support students' attempts at inquiry make more effective use of laboratory experiences than is typical. From these efforts the committee adduced four design principles to guide the development and implementation of effective laboratory experiences.

This chapter presents these design principles, their rationales and examples of their instantiation. It draws heavily on *ALR* (NRC, 2005), especially as that report synthesizes research on typical laboratories and more recent inquiry-oriented design research. Yet, the ideas expressed here should not be taken to reflect the consensus view of the high school labs committee unless explicitly noted as such. A brief summary of the problems of typical laboratory instruction is presented to support the design principles. This summary is drawn directly from *ALR*, to which readers may turn for a thorough review of the history of laboratory instruction in American high schools and the typical experiences of most current students. This is followed by a detailed description of the design principles derived in *ALR*. The scope of the study committee's work was constrained to high school science, so the discussion here is expanded to elaborate on research from elementary and secondary science education to provide a more thorough rationale for each design principle than is provided in *ALR*. The chapter closes with examples of how the design principles can be implemented to achieve many of the learning goals identified for labs in *ALR*, and which goals demand more research.

PROBLEMS OF TYPICAL LABORATORY INSTRUCTION

The most striking finding from *ALR* is that despite the longstanding interest in and centrality of laboratories in science education, the research base on labs is astonishingly weak. This problem arises from a number of factors. One factor is that arguments for laboratories hold out a number of goals that labs are meant to achieve. The committee identified goals that could be classified into seven areas: mastery of subject matter, developing scientific reasoning, understanding the complexity and ambiguity of empirical work, developing practical skills, understanding the nature of science, cultivating interest in science and learning science, and developing teamwork abilities (NRC, 2005). Research has not always assessed the achievement of these goals through laboratories in clear ways. Major reviews of laboratory research from the 1960s and 70s (Hofstein & Lunetta, 1982), and throughout the 1980s and 90s (Hofstein & Lunetta, 2004; Lazarowitz & Tamir, 1994) point to a number of flaws in this research that make it difficult to draw definitive conclusions. Earlier studies tended to provide poor descriptions of the constructs they were trying to assess and the methods that they used (a weakness pointed out by Hofstein & Lunetta, 1982). More recent studies tend to address these problems, but leave out important information about the nature of the laboratory activities studied, and particularly the roles that teachers played—issues that persist from earlier studies. Consequently, analyses of how particular forms of laboratory activities may or may not support specific pedagogic goals are scant. Studies comparing typical laboratory experiences with various forms of demonstrations (with or without discussion, for example) show no clear benefits to learning subject matter. Typical labs appear to produce slight improvements in aspects of students' scientific reasoning, and also in their interest in science (NRC, 2005). Looking at the overall body of research on typical laboratories over the last several decades, however, it appears that they fail to meet many of their purported goals (NRC, 2005). This stands in contrast to findings from research on what the committee labeled "integrated instructional units" that have more consistently positive findings and from which the lab design principles were drawn.

While research on the outcomes of laboratories is hard to read for definitive conclusions, findings are clearer on the problems with typical laboratory instruction. The research reviewed in *ALR* shows that students generally fail to perceive the purpose of the laboratory activities that they are asked to perform. A number of studies show that most students perceive the purpose of doing labs as "following the instructions" or "getting the right answer" (Hofstein & Lunetta, 2004, p. 38). This is true across all grade levels, from elementary school to college. Demonstrations and laboratories are often used in the classroom to produce a certain outcome or effect, and the scientific ideas that explain the outcome are often unclear (Reif, 1990). In high school, which was the focus of *ALR*, students tend to have labs about once per week. The overwhelmingly typical experience of American high school students is to perform "cookbook" labs, where the lab procedures are entirely spelled out. Such activities are the barest shadow of authentic scientific activity. Cookbook labs preclude any need for student effort to consider what they are trying to find out, or how they might do so.

The picture of the typical lab that is painted by available research is not pretty. Labs tend to be presented as things for students to do, with little attention to the purpose of the activity for student growth. Laboratory activities themselves too often focus on the manipulation of materials or apparatuses for no greater purpose, or they focus on calculations or simple representations of data without clear ideas of why such calculations are performed or representations created. This lack of purpose, unfortunately, may be a general feature of science instruction.

DESIGN PRINCIPLES FOR EFFECTIVE LABORATORY EXPERIENCES

In contrast to research on typical laboratory instruction, a growing body of research in science education has, over the last 20 years, focused on the design of inquiry-oriented learning environments. These approaches tend to come from the emergent field of the learning sciences, a field derived from cognitive science and borrowing liberally from anthropological studies of education. Learning sciences researchers approach the development of instruction from analyses of learning, commonly anchored in cognitive developmental research. That is, the forms and sequence of instruction are derived from analyses of both the competencies and difficulties that children of a particular age display when learning particular topics. Research on the interventions designed within this paradigm commonly focus on connecting processes of implementation to observed outcomes, through a methodology known as design-based (or simply design) research (Cobb, Confrey, diSessa, Lehrer, & Schauble, 2003; Design-Based Research Collective, 2003; Sandoval & Bell, 2004). Design-based research studies attracted the attention of the high school labs committee because these studies often showed positive outcomes on many of the learning goals articulated for laboratory instruction.

The important point from this area of research is that the crucial feature of laboratory activities is that they must be part of a coherent unit of instruction. There is no such thing as a good laboratory activity without reference to where that activity fits within some broader sequence of instruction. This point cannot be overstated. Teachers are inundated with activity options from professional development and teacher-oriented journals with very little information about the appropriateness of specific activities for instruction.

In reading the design principles that follow, readers may be struck by how obvious they seem. The review by *ALR* makes it clear that these design principles are not obvious. There are a number of reasons why this might be so. A look at any standards document for science education will show them to be largely unordered lists of facts and ideas that students should learn. Even within specific disciplines, there is tremendous variability in the extent to which core disciplinary ideas are seen as building upon one another. If the standards do not present coherent conceptual progressions, then why should the average curriculum be expected to do so? Teachers are not usually trained to develop coherent views of their subjects, but learn to organize specific lessons and manage instruction.

The following four design principles are proposed in *ALR* to reflect aspects of effective laboratory experiences as indicated by recent research into inquiry-oriented instructional efforts. The descriptions of each principle in *ALR* are quite brief. The descriptions here are expanded to provide the links to research that provide the rationale for each principle. These rationales are grounded in cognitive terms; that is, in the ways in which they can be expected to contribute to student reasoning through laboratory activities. There are other rationales possible, such as motivational, social, and institutional, but these factors are largely omitted here (and in *ALR*). Unless specifically noted as coming from *ALR*, these detailed rationales should not be interpreted as the consensus opinion of the high school labs study committee or the NRC.

Clearly Communicated Purposes

Effective laboratory experiences have clear learning goals that guide the design of the experience. Ideally these goals are clearly communicated to students. (NRC, 2005, p. 101)

It should go without saying that a good lab must have clear goals. It may very well be that many laboratory activities have clear goals, but as noted above most students do not comprehend them. It is not enough for a curriculum developer or teacher to have a clear goal, or set of goals, in mind for a laboratory activity. That goal must also be clear to students; students need to be informed of the goal to be given the chance to adopt that goal. The reasons for this idea are both cognitive and motivational.

Cognitive research shows that how the goals of some investigative activity get framed for learners affects their reasoning and performance. Dunbar (1993) compared two groups of college students' performance in using a computer microworld to determine gene function. One group was told to find evidence to confirm a given hypothesis, and the other group was told to come up with its own explanation. The explanation group was more thorough in their experimentation on the computer, and also more systematic and more careful in controlling experimental variables. Other cognitive researchers have described how children commonly perceive the purpose of experimentation in school to be generating some specific outcome or effect or making something happen rather than finding something out (Reif & Larkin, 1991; Schauble, Glaser, Duschl, Schulze, & John, 1995). Through middle and high school, many students report scientists conduct experiments not necessarily to test ideas but to see if their ideas are "right" (Carey, Evans, Honda, Jay, & Unger, 1989; Sandoval & Morrison, 2003). That is, the reason to do an experiment is to get a particular outcome, a "right" one.

Developmental research indicates that even a child as young as 5 years old can differentiate a hypothesis to test from the evidence that might confirm or refute it (Ruffman, Perner, Olson, & Doherty, 1993; Sodian, Zaitchik, & Carey, 1991; Tschirgi, 1980). Elementary aged children have productive ideas about how to deal with issues of uncertainty in their own investigations (Metz, 2004). With instruction on the purpose of experiments as tests of causal relations, students become better at designing controlled experiments (Schauble et al., 1995). Indeed, the developmental literature can be read as showing that students have productive capabilities for reasoning scientifically, but lack knowledge and experience (Lehrer & Schauble, 2006; Zimmerman, 2000).

Careful Sequencing Within the Flow of Instruction

Effective laboratory experiences are thoughtfully sequenced into the flow of classroom science instruction. That is, they are explicitly linked to what has come before and what will come after. (NRC, 2005, p. 102)

It seems ridiculous to state a principle of instruction that appears totally obvious. Yet, as described above, one of the main problems with typical laboratory experiences is that they are too often isolated from the rest of instruction. This principle extends the first one, in that not only do the immediate purposes of a lab experience need to be clear to students, how that experience builds upon previous lessons and where it leads also need to be made explicit. There are thus two aspects to this principle.

The first aspect is each laboratory experience should have a clear place within a carefully constructed sequence of instruction. The second is the ways in which the experience is linked to that sequence should be communicated to students.

The cognitive rationale underlying this principle derives from constructivist perspectives on learning, the idea that people construct knowledge from experience (Bransford, Brown, & Cocking, 1999). Central to contemporary views of learning is people build new knowledge from prior knowledge. Consequently, coherent units of instruction must begin with the elicitation of the range of student ideas and proceed to developing shared experiences and understanding that can then be built upon toward normative scientific understanding. Instruction must begin where students are in order to lead them to where we would like them to be. Laboratory experiences then, like any other kind of lesson, have to be considered for where in an instructional progression they might occur. The purpose of a particular lab activity could change depending upon its placement in the sequence. In the early part of an instructional unit, a good lab might be one that raises a number of questions in students minds, rather than a lab always being an activity in which students only answer written questions.

Integrated Focus on Concepts and Practice

[C]onceptual understanding, scientific reasoning, and practical skills are three capabilities that are not mutually exclusive. An educational program that separates the teaching and learning of content from the teaching and learning of processes is likely to be ineffective. (NRC, 2005, p. 102)

For a very long time there has been a divide between science content and science process in conceptualizations of science instruction. This separation is driven by the idea that there are methods of science, even a single "scientific method," independent of any specific scientific ideas, and such a method or methods can be taught to students in a generalized way. Both of these ideas are simply wrong.

Philosophical, historical, and sociological studies of scientific practice demonstrate there is no single scientific method, or even necessarily anything like it. Science is not a unified enterprise; rather, disciplines are comprised of collections of fields that develop their own methods to deal with the discipline's specific type and range of questions (Feyerabend, 1975/1993). Feyerabend, building on Kuhn's seminal work, argued that scientific methodologies were always historically contingent, as well as influenced by available technologies. Methods develop as responses to trying to solve particular problems of research. Studies of scientific laboratories support the idea that scientists perforce figure things out as they go, rather than following some prescribed method (Dunbar, 1995; Latour & Woolgar, 1986). Sociological studies of scientists highlight the social aspects of scientific activity, that is, of method, as implicit and explicit power relations influence the negotiated "construction of facts" (Latour, 1987; Latour & Woolgar, 1986). This is not to claim scientific "facts" are nothing more than social fabrications. Giere (1987), for example, points out putative facts must fit with observable reality. He summarizes contemporary views of science, however, in the rejection of some cold, logical rationalism as being the basis for scientific activity.

Taken as a whole, contemporary studies of scientific practice make clear not only that uniformity of method in the natural sciences does not exist but also that it is not even a goal. Rather, scientific communities strive to achieve standards for knowledge evaluation (Feyerabend, 1975/1993). Inherent in this formulation is the idea that the methods used to evaluate knowledge claims are inseparable from the claims themselves. Scientific processes are not separable from the scientific ideas that they generate. Theory and method are dialectically related: theory drives method, and empirical results force theory revision. The roots of this dialectical view extend at least to Kuhn's (1970) treatise on the nature of "scientific revolution," one of the key ideas of which is methods of investigation are heavily theory laden. The philosophical debates that solidified and extended this view are well summarized by Driver and colleagues (Driver, Leach, Millar, & Scott, 1996). Even before Kuhn's seminal work, Schwab's (1962) argument for science education organized through inquiry was the divorce of scientific processes from the ideas they generated was a mischaracterization of science as a "rhetoric of conclusions" (NRC, 2005, p. 26).

Another rationale for separating process and content in science instruction, and laboratory experiences specifically, is the idea that students need to learn certain general skills before they can productively investigate complex scientific ideas, or just that it is more efficient to teach "experimentation skills" generally. This is related to a view of "scientific thinking" that has been framed by cognitive scientists as a general skill divorced from knowledge in specific domains. Despite the continued popularity of this view in some quarters (Klahr & Simon, 1999; Kuhn, 2005), the overwhelming conclusion from research on learning over the last 50 or more years is people simply do not learn this way (Bransford et al., 1999; Lehrer & Schauble, 2006; NRC, 2007). Domain-specific knowledge, specifically knowledge about possible causal mechanisms, highly influences students' attempts at experimentation (Lehrer & Schauble, 2006; Zimmerman, 2000). The conclusion is that good laboratory experiences marry processes to the ideas those processes are intended to elaborate.

Ongoing Discussion and Reflection

Laboratory experiences are more likely to be effective when they focus students more on discussing the activities they have done during their laboratory experiences and reflecting on the meaning they can make from them, than on the laboratory activities themselves. (NRC, 2005, p. 102)

This principle may seem less obvious than the others. Evidence suggests that doing labs, even well designed ones, is insufficient to ensure that students learn what was intended. Thus, one reason for discussing labs is to check for student understanding. Public discussions about what students did, what they found out, how their findings and interpretations differ, and so on provide a needed mechanism for students to compare their understanding with those of their peers, and an opportunity to learn from more capable peers. Another reason is situating laboratory activities within ongoing efforts to make meaning from those activities is, as implied above, a more authentic practice of science (one that is feasible within science classrooms).

Cognitively, this principle is grounded partly in models of conceptual change (Posner, Strike, Hewson, & Gertzog, 1982; Strike & Posner, 1992). The theory of conceptual change posits that students must first become aware that their current conceptions are inadequate and need to be changed. Students' ideas are often tacit, however, and need to be made explicit before students will recognize discrepancies between their ideas and what they see. A number of researchers have implemented variations on the predict-observe-explain (White & Gunstone, 1992) cycle, wherein students make predictions or hypotheses about what is likely to happen in some situation, observe what does happen, and then explain the discrepancy. This cycle has been used to frame demonstrations and the discussions around them (Hatano & Inagaki, 1991; Minstrell & Stimpson, 1996), as well as to frame cycles of investigative activity within larger instructional units (Linn & Hsi, 2000; White, 1993). This approach suggests the importance of having students publicly articulate their ideas prior to laboratory experiences and to be required to reconcile those ideas with their lab results.

A corollary to this principle is that the laboratory experiences that are being discussed have to be meaningful to students. This point has been made already, but it bears repeating because it is so easy for this principle to be improperly applied. That is, it is easy to imagine students being asked to make predictions about an outcome that has nothing to do with conceptual ideas, and to putatively reflect on their experiences through lab report prompts that do not, in fact, demand much thought. The way around this is to see this principle as being one of the chief mechanisms for satisfying the principle of coherent sequencing. Appropriate discussion and reflection can provide the linkages between current lab experiences and what has come before and will come after.

Besides the cognitive benefits that accrue from demanding that students express and confront their own ideas, framing laboratory activities as legitimate and central objects of discussion in the science classroom can potentially address an as-yet elusive goal for science education—the development of students' ideas about the epistemological bases of science. Students commonly hold a view that scientists do experiments to get the "right" answer (Carey et al., 1989; Driver et al., 1996; Leach, 1999; Sandoval & Morrison, 2003), a view they have about their own school laboratory experiences. The well-established finding that students leave school with naïve ideas about the nature of science (Lederman, 1992; Sandoval, 2005) is a consequence of science instruction that presents scientific knowledge as a collection of "right" answers and labs as trivial activities to confirm them (Duschl, 1990; Lemke, 1990). Even the integrated instructional units reviewed by *ALR* have so far failed to show consistent development in students' epistemological ideas about science, but those efforts with some success indicate that explicit discussion of epistemological commitments, what we know and how, seems necessary (Sandoval, 2005).

To summarize, the high school labs committee was able to extract these four design principles through a review of recent research on inquiry-oriented interventions the committee labeled as integrated instructional units. As the name implies, these efforts integrate laboratory activities into a coherent sequence of instruction.

MEETING LEARNING GOALS THROUGH THE DESIGN PRINCIPLES

What does it look like to successfully enact these principles? *ALR* summarizes three examples of integrated instructional units the committee felt embodied the learning principles presented above: Chemistry That Applies (Lynch, Kuipers, Pyke, & Szesze, 2005), ThinkerTools (White, 1993; White & Frederiksen, 1998), and Computer as Learning Partner (Linn & Hsi, 2000). All three of these projects have developed integrated instructional units spanning several weeks of instruction, and *ALR* points out that each project shows evidence of achieving important learning goals proposed for laboratory instruction.

This section expands the summaries of these projects found in *ALR* and includes other successful integrated instructional units to describe how the design principles can be reified in curriculum and the evidence that positive results stem from the implementation of these principles. For illustrative purposes, projects have been selected across elementary and secondary science education, in contrast to the limited charge of *ALR* to examine high school science.

It will become apparent, as *ALR* indicates, that available research is variable in the goals addressed, with some goals receiving most of the attention, and others receiving little or none. The following learning goals are drawn directly from *ALR* (NRC, 2005, p. 196). One of the primary motives behind integrated instructional units has been to improve students' learning of the scientific concepts and theories within particular disciplines. There is copious evidence students' intuitive ideas about many natural phenomena are at odds with normative scientific conceptions, and these naïve ideas are quite difficult to change (Wandersee, Mintzes, & Novak, 1994). At the same time, most of these interventions have pursued the concurrent goal of developing scientific reasoning.

Enhancing Mastery of Subject Matter

Considerable evidence has been amassed over the last 2 decades that integrated instructional units lead to improved subject matter mastery compared to typical instruction, across a number of grade levels. In a new NRC report (NRC, 2007), much of what is known about elementary aged children's abilities to learn science and effective instructional methods is reviewed, and successful examples are provided at the middle school level, with fewer at the high school level.

An extremely successful integrated instructional unit highlighted in *ALR* is the ThinkerTools project (White, 1993; White & Frederiksen, 1998). ThinkerTools was designed as a 2 month curriculum to teach principles of Newtonian mechanics to sixth graders. Part of the motivation was to show that children this young are developmentally capable of reasoning abstractly (White, 1993). More germane to the present focus, however, was the intent to show that an alternative approach to instruction on this topic would improve students' conceptual understanding. ThinkerTools is designed to guide students through a sequence of models of force and motion that start from simple 1-dimensional problems and progress through more complex 2-dimensional problems and ultimately include friction forces. This sequence starts where it does based on cognitive research on students' ideas about force and motion, and the progression is organized according to an idea of the role that "intermediate" causal models play in scaffolding student learning (White, 1993). This approach emphasizes a coherent conceptual progression through the curriculum. Each module includes two kinds of laboratory activity that serve different purposes. Physical experimentation with objects, in the classroom, is used at the beginning of each module to provide a common phenomenal experience that students can become collectively engaged in explaining. Computer-based simulations are used to enable students to propose potential laws of motion and then test those laws through various motion simulations. The purpose of each of these laboratory activities is communicated to students though lab guides and teacher instructions that indicate why they are doing the activity and what they should focus on while doing them. Specific experiments, in the classroom or on the computer, are embedded within an extended stream of discussion about what the class is trying to find out and what they have learned. That is, a central component of the instructional approach is to emphasize discussion and reflection of laboratory activities. Results of classroom trials show that sixth graders outperform high school physics students on problems of force and motion (White, 1993), and that the inclusion of reflection activities is especially beneficial for low-achieving students (White & Frederiksen, 1998).

ThinkerTools is just one example of instructional efforts that base instruction on cognitive research on student conceptions, and use the careful sequencing of laboratory experiences to organize not just student work but student discussion and reflection on that work. The project-based science curricula developed at the University of Michigan have produced learning gains that exceed national samples on standardized tests and are particularly effective at producing gains on higher-level problems requiring the analysis and explanation of data (Rivet & Krajcik, 2004; Schneider, Krajcik, Marx, & Soloway, 2002).

Chemistry That Applies (*CTA*, State of Michigan, 1993) is another curriculum highlighted by the high school labs committee in *ALR* that follows the design principles and produces significant learning gains in the targeted concepts, related to the conservation of matter, when compared to alternatives (Lynch et al., 2005). In fact, Lynch and colleagues found while approximately one-fifth of the students in their sample "got" targeted concepts regardless of their curriculum, lower achieving students did better with *CTA* than with other instructional materials. This trend is found throughout the studies mentioned here and in others (Songer, Lee, & McDonald, 2003), and is most likely due to the focus on conceptual coherence that follows from the design principle of coherent sequence.

Developing Scientific Reasoning

Many of the interventions described above also lead to improvements in key aspects of scientific reasoning. ThinkerTools has been effective in helping students learn inquiry processes, especially processes of experimental design and interpretation (White & Frederiksen, 1998). White and Frederiksen show evidence these gains are, again, especially seen in lower-achieving students and stem from the design of specific prompts for reflection on aspects of scientific investigation. Similarly, the most striking achievements made by students in the project-based science interventions come on open-ended problems that require the evaluation of data to solve (Rivet & Krajcik, 2004; Schneider et al., 2002). These findings suggest students are not only learning "concepts" but also ways of reasoning about the natural world and representations of that world (i.e., data).

The design principles of using labs to develop content and process and to surround them with discussion and reflection appear to be largely responsible for gains in students' scientific reasoning. In their elementary grade modeling curriculum, Lehrer and Schauble (2006) have shown, over the course of many years, how careful sequencing of modeling activities develops children's abilities to represent data and build scientific models. Exemplifying the principle of coherent sequencing, Lehrer and Schauble develop laboratory activities that build off students' existing competence in drawing and physically modeling and move toward increasingly mathematic representations of data and trends. Their implementation of the principle of clear purpose is to enlist students in developing the purpose of activities, an approach pioneered effectively in the Fostering Communities of Learners program (Brown & Campione, 1990, 1994). In these interventions, students are engaged in processes of experimentation, representation, and analysis as means to develop conceptual understanding. That is, like professional scientific practice, methods are pursued to generate knowledge.

One caveat to these encouraging findings is the range of scientific reasoning targeted by these interventions has been relatively narrow. Chinn and Malhotra (2002) identified eleven aspects of "epistemological authenticity" they used to evaluate a number of research-based curricula. They found while these were much better than textbooks, they focused on a narrow band of scientific processes associated with the collection and interpretation of data. Chinn and Malhotra suggest the interventions they analyzed do a good job of capturing "one central epistemological feature of science—building and revising theoretical models" (Chinn & Malhotra, 2002, p. 204), but other important aspects of scientific reasoning are omitted, including posing research questions, considering methodological flaws, or examining the theory-ladenness of observation.

Understanding the Nature of Science

In general, inquiry-oriented reform efforts, including what *ALR* describes as integrated instructional units, have not been especially successful in helping students to develop more sophisticated epistemological ideas about science, except a few cases where epistemological ideas have been an explicit focus of instruction over an extended period of time (Sandoval, 2005). The state of research currently seems to be both a lack of impact and a lack of attention. None of the integrated instructional units described so far in this section have held the development of nature of science (NOS) ideas as a primary goal. The focus has been on getting students to learn how to conduct inquiry, as opposed to reflecting on the nature of that inquiry as one kind of knowledge-making enterprise. A few projects have examined students' ideas about NOS with limited findings.

The Computer as Learning Partner project (Linn, 1991), which later became the Knowledge Integration Environment (Linn, Bell, & Hsi, 1998), and then the Web-based inquiry science environment (Linn & Hsi, 2000) took an early interest in looking at students' ideas about science and how they might change through integrated instruction. Consistent with other studies (Sandoval, 2005), they found most students had a range of inconsistent ideas about science, those ideas did not change with instruction, but those students with more sophisticated ideas appeared to learn more from their curriculum (Davis, 2003; Linn & Songer, 1993).

The BGuILE project is another example where epistemological development was a target of instruction but not only was no change seen in students' ideas, their reported ideas were highly inconsistent (Sandoval & Morrison, 2003). In this case, instructional activities focused on the development of causal explanations for evolutionary change that required students to coordinate data with a Darwinian theoretical framework. As with other integrated instructional units, students showed they could engage successfully in epistemologically authentic scientific practices, but they did not see such work as related to the work of scientists.

Explicit discussion of the epistemological aspects of inquiry, including laboratory experiences, appears to help at least some students develop more sophisticated ideas about the nature of science (Khishfe & Abd-El-Khalick, 2002; Khishfe & Lederman, 2006). A study by Smith and her colleagues (Smith, Maclin, Houghton, & Hennessey, 2000) is interesting because of its singularity: students had the same science teacher from the first through sixth grades, and their science instruction was organized

entirely around extended investigations in which the children took leading roles in the topics studied and the methods to investigate them–an
approach that lends itself to the effective implementation of the first
three design principles. Lab purposes are clear to students because the
students generate them; process and content are naturally combined as
students investigate problems of their own design, and coherence of units
is aided because the units are built around questions of interest. This
study is also interesting because issues of epistemological importance
were emphasized in the teacher's approach: students were consistently
asked to make hypotheses, justify experimental designs, interpret data,
and challenge claims made from data. This extended reflection upon
one's own practices of inquiry helped these children to develop sophisticated ideas about the nature of scientific practice.

These studies implicate the principle of ongoing discussion and reflection as key to the development of epistemological ideas. Yet, these studies
fail to describe in detail the structure or content of these epistemologically oriented discussions. Consequently, they provide little guidance
about how to implement this design principle. Ongoing interest in the
place of argumentation in science instruction and its potential to highlight epistemological aspects of science may improve this picture,
although it remains difficult to link student practice to underlying epistemological ideas (Sandoval & Millwood, 2008).

Cultivating Interest in Science

Interest, motivation, and other affective outcomes have received little
attention in research involving integrated instructional units. Both
ThinkerTools and *CTA* are highlighted in *ALR* for their approaches to
measuring student engagement in each curriculum. In one ThinkerTools
study, students were more likely to turn in final project reports (about
projects that small groups designed and conducted) than comparison
groups, and individual reports were less likely to be identical to a
partner's work (NRC, 2005; White & Frederiksen, 1998). White and
Frederiksen used these results as indicators of students' engagement in
curricular tasks. In this same study, White and Frederiksen also found that
after the curriculum, fewer students reported that being good at science
was innate, and fewer agreed that boys were better at science than girls
(see NRC, 2005, p. 98). In a large-scale study of *CTA*, students using that
curriculum showed a higher level of participation in activities than did
students using comparison curricula and were more likely to pursue deep
learning goals than comparison students (Lynch et al., 2005).

ALR points out, however, direct assessments of attitude, motivation, and interest are missing from the research reviewed by the committee. A recent review of research within the learning sciences, from which much of the integrated instructional units come, indicates a failure to connect this work with the longstanding research on motivation and points out the need to make assessments of motivation and interest a more explicit focus of research on integrated instructional units (Blumenfeld, Kempler, & Krajcik, 2006).

Evidence on Other Goals

As indicated earlier, the high school labs committee identified three other goals for laboratory instruction: understanding the complexity and ambiguity of empirical work, developing practical skills, and developing teamwork skills. The committee found no substantial evidence that integrated instructional units met these goals or even addressed them (NRC, 2005). Although *ALR* does not speculate on why this is, the intellectual approach to science learning and instruction reflected in research on integrated instructional units may minimize these goals. This research has focused primarily on developing students' scientific reasoning, often in service of conceptual understanding. A common approach is to frame instruction as a cognitive apprenticeship (Collins, Brown, & Newman, 1989) focused on "engaging students in the reasoning and discursive practices of scientists, and not necessarily the exact activities of professional scientists" (Sandoval & Reiser, 2004, p. 347). This perspective minimizes the value of assessing students' acquisition of practical laboratory skills, as those skills are not seen as especially valuable in themselves. Arguably, assessments of teamwork suffer the same fate, as teamwork, per se, acquires importance only with respect to how it enables or constrains productive collaboration. That is, the learning sciences perspective underlying much of the work identified by the committee as integrated instructional units does not hold practical skills or teamwork as end goals. Consequently, they tend not to be assessed.

With respect to the goal of understanding methodological complexity and ambiguity, this may be a blind spot in conceptualizations of the nature of science. *ALR* conceptualizes this goal as having to do with the ways in which scientists attempt to deal with issues such as measurement error and data aggregation and representation (NRC, 2005). Research on students' understanding of empirical methods in science has focused almost exclusively on experimental control (Sandoval, 2005), and research-based inquiry tasks tend not to address ambiguity (Chinn & Malhotra, 2002). On the other hand, the concerns reflected in this goal are quite similar to the goals pursued in Lehrer and Schauble's research on "data modeling" (Lehrer & Romberg, 1996; Lehrer & Schauble, 2004), including how one decides what data to get, how to get them, how to represent them, how to deal with measurement error and variability, and so forth. Interestingly, Lehrer and Schauble have pursued this work with elementary aged children with some success. Other recent evidence suggests children can productively reason about sources of uncertainty in investigations of their own design, and strategies for dealing with them, as early as second grade (Metz, 2004). As with the research on other aspects of scientific epistemology discussed above, this work suggests the primacy of designing appropriate opportunities for ongoing discussion and reflection, and for explicitly connecting investigative methods to the knowledge they are trying to produce or evaluate.

CONCLUSION

America's Lab Report describes the situation of typical laboratory experiences in American high schools as in need of serious attention and repair. The report urges research into laboratory instruction that can make high quality laboratory experiences common for all students. Addressing this challenge requires researchers to more clearly articulate the roles that specific laboratory activities play in meeting specific learning goals. The challenge "is given greater urgency in view of the multiple pressures placed on schools and districts to increase the performance of a diverse student body" (NRC, 2005, p. 199). These pressures are social, economic, and pedagogical. The design principles proposed in *ALR* and expanded upon here provide a basis for addressing pedagogical challenges. *ALR* points out the research on integrated instructional units has not yet achieved broad scale, nor have researchers developed effective means for distributing knowledge about effective practices to the audiences that most need them: teachers and local education agencies and personnel.

Only a few of the research efforts described here have been studied at any sort of scale, with 1,000 or more students (Lynch et al., 2005; Schneider et al., 2002; Songer et al., 2003). One of the critical needs of large-scale research is to devise methods to account for the effects from contextual variations on the implementation of specific curricula, including variables at individual, classroom, school, and community levels. *ALR* points out the need for such research, but how to do it remains an open question. It may require research partnerships that draw on the expertise and methods of quite disparate fields and traditions that can, for example, construct coherent accounts across psychological and sociological levels. This kind of collaboration remains challenging to conceive and execute.

The value in the design principles described here will come from research efforts that can use them to propose particular roles for laboratory activities and the discourse surrounding them and how those roles achieve specific sets of instructional goals. Notice the exemplary programs highlighted in *ALR* address multiple goals at the same time through particular activities. Notice also the *ALR* principles are effective because activities are embedded within forms of discourse effectively transforming the ways of doing and talking science in classrooms. What is studied in these interventions are not particular activities, or even necessarily curriculum, but classroom ecologies. Examining the role of kinds of laboratory activities makes sense only in terms of analyses of their place within specific ecologies. Conversely, understanding how to achieve quality science instruction, including quality lab experiences, will come from research that can specify variations in classroom ecology affecting enactments of particular criteria. We already know many of the attributes of effective instructional materials. What we need to know is how these attributes are sensitive to differences between students, teachers, and schools.

ACKNOWLEDGMENTS

I am grateful to Jean Moon of the Board on Science Education at the National Research Council for the opportunity to serve on the high school labs study committee. I thank Margaret Hilton and Susan Singer for comments on this chapter. The views and opinions expressed here are mine only, and do not necessarily reflect the views of the high school labs study committee or the National Research Council.

REFERENCES

Blumenfeld, P. C., Kempler, T. M., & Krajcik, J. S. (2006). Motivation and cognitive engagement in learning environments. In R. K. Sawyer (Ed.), *The Cambridge handbook of the learning sciences* (pp. 475-488). New York: Cambridge University Press.

Bransford, J. D., Brown, A. L., & Cocking, R. (1999). *How people learn: brain, mind, experience, and school*. Washington, DC: National Academy Press.

Brown, A. L., & Campione, J. C. (1990). Communities of learning and thinking, or a context by any other name. *Human Development, 21*, 108-126.

Brown, A. L., & Campione, J. C. (1994). Guided discovery in a community of learners. In K. McGilly (Ed.), *Classroom lessons: Integrating cognitive theory and classroom practice* (pp. 229-270). Cambridge, MA: MIT Press.

Carey, S., Evans, R., Honda, M., Jay, E., & Unger, C. (1989). "An experiment is when you try it and see if it works": A study of Grade 7 students' understanding of the construction of scientific knowledge. *International Journal of Science Education, 11*, 514-529.

Chinn, C. A., & Malhotra, B. A. (2002). Epistemologically authentic inquiry in schools: a theoretical framework for evaluating inquiry tasks. *Science Education, 86*, 175-218.

Cobb, P., Confrey, J., diSessa, A., Lehrer, R., & Schauble, L. (2003). Design experiments in educational research. *Educational Researcher, 32*(1), 9-13.

Collins, A., Brown, J. S., & Newman, S. E. (1989). Cognitive apprenticeship: Teaching the crafts of reading, writing, and mathematics. In L. B. Resnick (Ed.), *Knowing, learning, and instruction: Essays in honor of Robert Glaser* (pp. 453-494). Hillsdale, NJ: Erlbaum.

Davis, E. A. (2003). Untangling dimensions of middle school students' beliefs about scientific knowledge and science learning. *International Journal of Science Education, 25*(4), 439-468.

Design-Based Research Collective. (2003). Design-based research: an emerging paradigm for educational inquiry. *Educational Researcher, 32*(1), 5-8.

Driver, R., Leach, J., Millar, R., & Scott, P. (1996). *Young people's images of science.* Buckingham, England: Open University Press.

Dunbar, K. (1993). Concept discovery in a scientific domain. *Cognitive Science, 17*, 397-434.

Dunbar, K. (1995). How scientists really reason: Scientific reasoning in real-world laboratories. In R. J. Sternberg & J. E. Davidson (Eds.), *The nature of insight* (pp. 365-395). Cambridge, MA: MIT Press.

Duschl, R. A. (1990). *Restructuring science education: the importance of theories and their development.* New York: Teachers College Press.

Feyerabend, P. (1993). *Against method* (3rd ed.). New York: Verso. (Original work published 1975)

Giere, R. N. (1987). The cognitive study of science. In N. J. Nersessian (Ed.), *The Process of Science* (pp. 139-159). Dordrecht: Martinus Nijhoff.

Hatano, G., & Inagaki, K. (1991). Sharing cognition through collective comprehension activity. In L. B. Resnick, J. M. Levine & S. D. Teasley (Eds.), *Perspectives on socially shared cognition* (pp. 331-348). Washington, DC: American Psychological Association.

Hofstein, A., & Lunetta, V. N. (1982). The role of the laboratory in science teaching: Neglected aspects of research. *Review of Educational Research, 52*(2), 201-217.

Hofstein, A., & Lunetta, V. N. (2004). The laboratory in science education: Foundations for the twenty-first century. *Science Education, 88*, 28-54.

Khishfe, R., & Abd-El-Khalick, F. (2002). Influence of explicit and reflective versus implicit inquiry-oriented instruction on sixth graders' views of nature of science. *Journal of Research in Science Teaching, 39*(7), 551-578.

Khishfe, R., & Lederman, N. G. (2006). Teaching nature of siene within a controversial topic: integrated versus nonintegrated. *Journal of Research in Science Teaching, 43*(4), 395-418.

Klahr, D., & Simon, H. A. (1999). Studies of scientific discovery: complementary approaches and convergent findings. *Psychological Bulletin, 125,* 524-543.

Kuhn, D. (2005). *Education for thinking.* Cambridge, MA: Harvard University Press.

Kuhn, T. S. (1970). *The structure of scientific revolutions* (2nd ed.). Chicago: University of Chicago Press.

Latour, B. (1987). *Science in action.* Cambridge, MA: Harvard University Press.

Latour, B., & Woolgar, S. (1986). *Laboratory life: the construction of scientific facts* (2nd ed.). Princeton, NJ: Princeton University Press.

Lazarowitz, R., & Tamir, P. (1994). Research on using laboratory instruction in science. In D. L. Gabel (Ed.), *Handbook of research on science teaching and learning* (pp. 94-130). New York: Macmillan.

Leach, J. (1999). Students' understanding of the coordination of theory and evidence in science. *International Journal of Science Education, 21*(8), 789-806.

Lederman, N. G. (1992). Students' and teachers' conceptions of the nature of science: a review of the research. *Journal of Research in Science Teaching, 29*(4), 331-359.

Lehrer, R., & Romberg, T. (1996). Exploring children's data modeling. *Cognition & Instruction, 14*(1), 69-108.

Lehrer, R., & Schauble, L. (2004). Modeling natural variation through distribution. *American Educational Research Journal, 41*(3), 635-679.

Lehrer, R., & Schauble, L. (2006). Scientific thinking and science literacy. In W. Damon, R. Lerner, K. A. Renninger & I. E. Sigel (Eds.), *Handbook of child psychology: Child Psychology in Practice* (6th ed., Vol. 4., pp. 153-196). Hoboken, NJ: Wiley.

Lemke, J. L. (1990). *Talking science: language, learning, and values.* Norwood, NJ: Ablex.

Linn, M. C. (1992). The computer as learning partner: Can computer tools teach science? In L. Roberts, K. Sheingold, & S. Malcolm (Eds.), *This year in school science 1991: Technology for teaching and learning.* Washington, DC: AAAS.

Linn, M. C., Bell, P., & Hsi, S. (1998). Using the internet to enhance student understanding of science: The knowledge integration environment. *Interactive Learning Environments, 6*(1-2), 4-38.

Linn, M. C., & Hsi, S. (2000). *Computers, teachers, and peers.* Mahwah, NJ: Erlbaum.

Linn, M. C., & Songer, N. B. (1993). How do students make sense of science? *Merrill-Palmer Quarterly, 39*(1), 47-73.

Lynch, S., Kuipers, J., Pyke, C., & Szesze, M. (2005). Examining the effects of a highly rated science curriculum unit on diverse students: results from a planning grant. *Journal of Research in Science Teaching, 42*(8), 912-946.

Metz, K. E. (2004). Children's understanding of scientific inquiry: their conceptualization of uncertainty in investigations of their own design. *Cognition and Instruction, 22*(2), 219-290.

Minstrell, J., & Stimpson, V. (1996). A classroom environment for learning: guiding students' reconstruction of understanding and reasoning. In L. Schauble

& R. Glaser (Eds.), *Innovations in learning: New environments for education* (pp. 175-202). Mahwah, NJ: Erlbaum.

National Research Council. (2005). *America's Lab Report: Investigations in high school science.* Washington, DC: National Academy Press.

National Research Council. (2007). *Taking science to school: Learning and teaching science in Grades K-8.* Washington, DC: National Academy Press.

Posner, G. J., Strike, K. A., Hewson, P. W., & Gertzog, W. A. (1982). Accommodation of a scientific conception: toward a theory of conceptual change. *Science Education, 66*(2), 211-227.

Reif, F. (1990). Transcending prevailing approaches to science education. In M. Gardner, J. G. Greeno, F. Reif, A. H. Schoenfeld, A. diSessa & E. Stage (Eds.), *Toward a Scientific Practice of Science Education* (pp. 91-109). Hillsdale, NJ: Erlbaum.

Reif, F., & Larkin, J. H. (1991). Cognition in Scientific and Everyday Domains: Comparison and Learning Implications. *Journal of Research in Science Teaching, 28*(9), 733-760.

Rivet, A. E., & Krajcik, J. S. (2004). Achieving standards in urban systemic reform: an example of a sixth grade project-based science curriculum. *Journal of Research in Science Teaching, 41*(7), 669-692.

Ruffman, T., Perner, J., Olson, D. R., & Doherty, M. (1993). Reflecting on scientific thinking: children's understanding of the hypothesis-evidence relation. *Child Development, 64*, 1617-1636.

Sandoval, W. A. (2005). Understanding students' practical epistemologies and their influence on learning through inquiry. *Science Education, 89*, 634-656.

Sandoval, W. A., & Bell, P. (Eds.). (2004). Design-based research methods for studying learning in context [Special Issue]. *Educational Psychologist, 39*(4).

Sandoval, W. A., & Millwood, K. A. (2008). What can argumentation tell us about epistemology? In S. Erduran & M. P. Jiménez-Aleixandre (Eds.), *Argumentation in science education: perspectives from classroom-based research* (pp. 68-85). New York: Springer.

Sandoval, W. A., & Morrison, K. (2003). High school students' ideas about theories and theory change after a biological inquiry unit. *Journal of Research in Science Teaching, 40*(4), 369-392.

Sandoval, W. A., & Reiser, B. J. (2004). Explanation-driven inquiry: integrating conceptual and epistemic supports for science inquiry. *Science Education, 88*, 345-372.

Schauble, L., Glaser, R., Duschl, R. A., Schulze, S., & John, J. (1995). Students' understanding of the objectives and procedures of experimentation in the science classroom. *Journal of the Learning Sciences, 4*(2), 131-166.

Schneider, R. M., Krajcik, J. S., Marx, R. W., & Soloway, E. (2002). Performance of students in project-based science classrooms on a national measure of science achievement. *Journal of Research in Science Teaching, 39*(5), 410-422.

Schwab, J. J. (1962). The teaching of science as enquiry. In J. J. Schwab & P. Brandwein (Eds.), *The Teaching of Science* (pp. 3-103). Cambridge, MA: Harvard University Press.

Smith, C. L., Maclin, D., Houghton, C., & Hennessey, M. G. (2000). Sixth-grade students' epistemologies of science: the impact of school science experiences on epistemological development. *Cognition & Instruction, 18*(3), 349-422.

Sodian, B., Zaitchik, D., & Carey, S. (1991). Young children's differentiation of hypothetical beliefs from evidence. *Child Development, 62,* 753-766.

Songer, N. B., Lee, H.-S., & McDonald, S. (2003). Research towards an expanded understanding of inquiry science beyond one idealized standard. *Science Education, 87,* 490-516.

State of Michigan. (1993). *Chemistry that applies.* Lansing, MI: Author.

Strike, K. A., & Posner, G. J. (1992). A revisionist theory of conceptual change. In R. A. Duschl & R. J. Hamilton (Eds.), *Philosophy of science, cognitive psychology, and educational theory and practice* (pp. 147-176). Albany, NY: SUNY Press.

Tschirgi, J. E. (1980). Sensible reasoning: A hypothesis about hypotheses. *Child Development, 51,* 1-10.

Wandersee, J. H., Mintzes, J. J., & Novak, J. D. (1994). Research on alternative conceptions in science. In D. L. Gabel (Ed.), *Handbook of research on science teaching and learning* (pp. 177-210). New York: Macmillan.

White, B. Y. (1993). ThinkerTools: Causal models, conceptual change, and science education. *Cognition and Instruction, 10*(1), 1-100.

White, B. Y., & Frederiksen, J. R. (1998). Inquiry, modeling, and metacognition: Making science accessible to all students. *Cognition & Instruction, 16*(1), 3-118.

White, R. T., & Gunstone, R. (1992). *Probing understanding.* London: Falmer Press.

Zimmerman, C. (2000). The development of scientific reasoning skills. *Developmental Review, 20,* 99-149.

CHAPTER 3

THE USE OF SPECIALIZED FACILITIES FOR LABORATORY SCIENCE INSTRUCTION IN ELEMENTARY SCHOOLS

Deborah L. Hanuscin

The *National Science Education Standards* emphasize teachers should design and manage learning environments that provide students with the time, space, and resources needed for learning science. At the elementary level, this "space" is typically a multipurpose classroom; however, increasingly elementary schools are investing personnel as well as financial and other resources into specialized laboratory facilities for science. How do separate facilities for science support and contribute to ensuring adequate and appropriate laboratory instruction at the elementary level? This chapter explores the challenges to implementing laboratory instruction at the elementary level and examines the role that specialized facilities might play in effective science teaching in the elementary grades.

INTRODUCTION

In February 2007, the National Science Teachers Association (NSTA) reissued a position statement in support of the integral role of laboratory

The Impact of the Laboratory and Technology on Learning and Teaching Science K-16, pp. 57–70

investigations in science instruction. Drawing upon the National Research Council's (2006) summary of research in *America's Lab Report*, NSTA defined a laboratory investigation or "lab" as "an experience in the laboratory, classroom, or the field that provides students with opportunities to interact directly with natural phenomena or with data collected by others using tools, materials, data collection techniques, and models" (National Research Council [NRC], 2006, p. 3). As indicated by this definition, however, "laboratory" can also be used to denote the physical space in which the experience takes place. At the secondary level, the science classroom is typically a laboratory space. At the elementary level, however, the science classroom is typically a multipurpose classroom in which all subjects are taught. Yet, increasingly, elementary schools are investing their resources in establishing specialized facilities for science that more closely resemble the "lab" rooms of middle and secondary schools. These may be used in conjunction with a specialist-led science program, or may simply be a way to pool resources for the benefit of all teachers. Understanding the role of laboratory activities in science learning can be enhanced by better understanding the context in which that learning takes place.

How do separate facilities for science support and contribute to ensuring adequate and appropriate laboratory instruction at the elementary level? This chapter begins by considering the challenges to teaching elementary science, and the way in which specialized laboratory facilities might assist schools in overcoming these barriers. Next, it includes a discussion of the purposes and uses of specialized elementary school science facilities and how those align with the goals of laboratory instruction, in the context of the existing research in this area. The chapter concludes with a description of a research agenda, that when enacted would investigate, using several current lines of research, the impact of elementary science laboratory facilities on effective science instruction.

CHALLENGES TO TEACHING ELEMENTARY SCIENCE

The *National Science Education Standards* (NRC, 1996) emphasize teachers should design and manage learning environments that provide students with the time, space, and resources needed for learning science. Despite these emphases, however, research reveals that relatively little class time— less than 30 minutes per day—is spent on science instruction in the elementary schools (Fulp, 2002). Comparison of results from the 1993 and 2000 *National Survey of Science and Mathematics Education* (Smith, Banilower, McMahon, & Weiss, 2002) suggest there has been little change in science instruction in the nation as a whole since the publication of the *National Science Education Standards*.

A variety of factors contribute to this problem. Most notably, elementary teachers' anxiety and negative attitudes about teaching science have been well documented in previous decades (Czerniak & Chiarelott, 1990; Westerback, 1982). More recent surveys indicate while 77% of elementary teachers consider themselves "well qualified" to teach language arts/reading, fewer than 3 in 10 indicate they feel "well qualified" to teach science (Weiss, Banilower, McMahon, & Smith, 2001). This leads to a lack of confidence and some teachers avoiding teaching science altogether (Tilgner, 1990). Those teachers who *do* feel confident, however, may still encounter additional barriers to teaching science; in addition to inadequate content background, two obstacles to teaching science most frequently cited by elementary teachers include a lack of science equipment and inadequate time and space for science instruction (Tilgner, 1990). More recent data from the 2000 *National Survey of Science and Mathematics Education* further substantiates this finding; along with lack of content preparation, inadequate facilities and equipment and lack of money to purchase consumable supplies were identified as barriers to the effective and equitable teaching of science (Banilower, Smith, & Weiss, 2002). Twenty percent of the schools in this survey cited facilities as a "serious problem" for science and mathematics instruction.

In order to ensure that laboratory investigations are implemented in schools, the NSTA emphasizes that adequate facilities are essential (NSTA, 2007). Yet, given the above, it stands to reason that a significant number of teachers lack the necessary facilities for science instruction. What do these teachers do in the absence of adequate facilities to successfully conduct laboratory investigations with their students? Articles in NSTA professional journals offer insight about strategies teachers and schools use to cope with less-than-adequate facilities for science instruction (e.g., Fehlig, 1996; Fox, 1994; Higle, 1984; Mackin & Williams, 1995). In particular, these provide suggestions for creating "elementary science labs" that resemble secondary science classrooms.

FACILITIES FOR ELEMENTARY SCIENCE INSTRUCTION

One major difference between elementary and middle or high school is the nature of the classroom. Typical elementary school classes are "self-contained," where a single teacher is responsible for teaching all or most of the academic subjects to a single group of students. Thus, science is taught in a multipurpose classroom, as opposed to specialized science laboratories, as is usually the case in middle and high schools. Yet, it is not unheard of for elementary schools to have separate laboratory facilities for science (Biehle, Motz, & West, 1999; Harbeck, 1985; Vorsino, 1992).

As far back as 1951, the NSTA's *School Facilities for Science Instruction* (Richardson, 1951) discussed the issue of whether to place the facilities for science in the regular classroom where most other activities are held or to use a separate room. According to the NSTA, the flexibility of the elementary classroom is key; there must be enough space and materials to accommodate a wide range of activities, and the "unique needs of science teaching should be anticipated in planning such general features as … illuminating, ventilating, plumbing, and electrical services" (p. 8).

In its 1951 guide (Richardson, 1951), the NSTA advocated for elementary science to be taught in the regular classroom "as far as possible" (p. 165), but acknowledge that because not all elementary classrooms may be equipped to meet the needs of science instruction, some activities must be provided elsewhere:

> While the separation of the facilities for experiences in science from the room where other activities are carried on has undesirable features, it is better to have such separated facilities than none at all. The planning of a new building should ordinarily provide for science facilities within the room. The conversion of existing facilities should likewise include provisions for science experiences within the elementary room, but in exceptional cases it may be necessary to use a separate room. (p. 165)

Yet, no further clarification of the "undesirable features" was given, leaving schools and districts with little other than local circumstances on which to base their decisions.

The more recent *NSTA Guide to School Science Facilities* (Biehle et al., 1999) extended the discussion of elementary science laboratories beyond simply the physical space to consider the way in which instruction of science in separate facilities might reinforce the separation of disciplines, versus integration of science and other academic subjects:

> The most basic decision at the elementary level is whether to build separate science facilities or to make all classrooms—or one of each shared pair of classrooms—science friendly. Most schools that have the resources opt to do both, putting the basics for science in every room, while creating a special space for more in-depth discovery. This approach facilitates science's integration with other subjects, while stressing the unique characteristics of the science environment. (p. 31)

In this manual, we are also provided a description of the elementary science laboratory:

> often called the "discovery room," [the lab] is a large, well-equipped, and well-lit classroom with sinks and extensive storage for science equipment and kits. These materials may be used in the room or checked out for

follow-up activities in the homeroom. While a discovery-room approach requires cooperation and schedule coordination among the elementary staff, most teachers feel that the advantages of this specialized science room outweigh any administrative requirements. (p. 32)

Rather than a discussion of the aforementioned advantages or administrative requirements, however, what follows this description is a set of guidelines and specifications for furnishings (including work space, storage, display space, and teacher space) as well as lighting, utilities, and computers. Clearly, the establishment of separate facilities for science instruction at the elementary level requires significant resources—yet no evidence is provided to support the idea that utilization of separate facilities for science at the elementary level is worth this investment. For that, we turn to the research literature.

LABORATORY INSTRUCTION IN THE ELEMENTARY SCHOOL

Laboratory instruction has played a central role in science education and has been recognized for its potential in promoting meaningful learning *if* students are given sufficient opportunity in an environment that supports knowledge construction (Tobin, 1990). Studies conducted over the past several decades suggest "laboratory work is an important medium for enhancing attitudes, stimulating interest and enjoyment, and motivating students to learn science" (Hofstein & Lunetta, 2004). The laboratory itself, according to Hofstein and Lunetta, is a "unique environment":

> The social environment in a school laboratory is usually less formal than in a conventional classroom; thus, the laboratory offers opportunities for productive, cooperative interactions among students and with the teacher that have the potential to promote an especially positive learning environment. (p. 35)

Research conducted on laboratory instruction has focused on the learning environment, which depends greatly on the nature of the activities conducted in the lab, students' perceptions of laboratory work, teacher expectations, and the nature of assessment. Laboratory instruction "is influenced, in part, by the materials, apparatus, resources, and physical setting" (Hofstein & Lunetta, 2004, p. 35). In the middle and secondary levels, specialized science classrooms are the norm; however, in elementary schools science may be taught in either a "conventional" multipurpose classroom *or* in a specialized laboratory facility.

While studies abound that examine laboratory instruction in schools, a review of the literature yields only three studies that use "laboratory" to denote specialized science facilities within elementary schools. One of these specifically examines the implementation and utilization of an elementary science laboratory for instruction by regular classroom teachers in a traditional delivery model (Vorsino, 1992), while two cite laboratory facilities as part of an investigation of delivery models for elementary science instruction that utilize science specialists (Jones & Edmunds, 2006; Schwartz, Abd-El-Khalick, & Lederman, 2000). A discussion of each of these follows.

A Science Lab for All Elementary Teachers

Vorsino (1992) examined the use of a separate science classroom facility at an elementary school, the purpose of which was to facilitate implementation of laboratory investigations. The lab was initiated by PTA members who persuaded local business leaders to donate a portable classroom building to house a science lab. Later that year, the school applied for and received a grant to purchase equipment, supplies, and materials for the lab. Because of district-wide budget cuts, the cost of hiring specialists to teach science was prohibitive. Therefore, rather than being designated as a room in which specialists would teach science, the laboratory was set up for use by classroom teachers and a schedule was created to allow each class to visit the laboratory several times per week.

According to Vorsino (1992), a variety of problems prevented successful implementation of the laboratory. These spanned a wide range of issues, and in some cases, the laboratory created new problems related to those it was designed to address—specifically, the availability of materials for science instruction. The science laboratory had been equipped with non-consumable materials; as Vorsino pointed out, the problem for many teachers became the expense of purchasing the consumable supplies required for many of the science laboratory activities (cf. Weiss et al., 2001). Even more so, finding sufficient storage in the laboratory to keep newly purchased items became a problem.

Other problems were logistical in nature (Vorsino, 1992). Though a lab schedule had been arranged, conflicts arose when special assemblies, programs, or field trips coincided with teachers' laboratory time slots. A full lab schedule made it difficult for those teachers to "make up" lost lab time, and often they simply resorted to using their own classrooms. Furthermore, teachers complained that the schedule provided insufficient time for preparation and clean-up. Other teachers believed travel time to and from the lab detracted from the instructional benefits of the lab. For example, the kindergarten classrooms were located across the school campus from the laboratory. Behavior problems, compounded by increased class size, occurred en route to and from the laboratory. Thus, instructional time was lost in the transition. As a result, kindergarten teachers decided not to use the laboratory, but rather conduct science activities in their regular classrooms.

Though each class was scheduled for a laboratory session, Vorsino (1992) noted the science laboratory was not being used on a regular basis. She inferred that frustration with securing a desirable time slot led to this lack of use; however, Vorsino also speculated the newness of the laboratory itself may have seemed daunting, and unfamiliarity with science laboratories may have caused teachers to avoid using it. Thus, barriers to effective science teaching mentioned previously (e.g., lack of science content preparation and pedagogical knowledge for science teaching) may still serve as barriers to teaching science within a specialized laboratory facility.

The following school year, a committee was formed to examine use of the laboratory. The committee, comprised of administrators, teachers, and parents, met on a bimonthly basis, and developed several strategies to address issues related to implementing the laboratory facility Parent and community volunteers were recruited to provide instruction and assist with set-up, clean-up, maintenance, and preparation. A minigrants program was created to provide funds to teachers to purchase consumable supplies. Professional development was provided to teachers to orient them to the laboratory, and demonstration lessons were used to provide teachers with models of effective laboratory instruction. By the end of the year, teachers were using the laboratory on a regular basis and several indicators provided evidence of the impact of the program; the number of students submitting projects for the science fair more than doubled, and student achievement in science had increased both on report cards and standardized tests.

According to Vorsino (1992), the science laboratory committee concluded with several ongoing recommendations to ensure successful implementation of the laboratory: (1) the curriculum should successfully integrate laboratory instruction with classroom experiences; (2) volunteers should be utilized to provide instruction and assistance to teachers in utilizing the laboratory; (3) assessment should be utilized to evaluate student progress and effectiveness laboratory instruction; (4) the lab should be centrally located and accessible to all students; and (5) a committee of stakeholders should regularly monitor and support use of the laboratory school-wide.

Science Laboratories for Elementary Science Specialists

Though there is still debate regarding delivery models for elementary science (see Gess-Newsome, 1999), it is believed by some science educators that employing specialists to teach science in appropriately equipped elementary science rooms would abate the constraints of priority, time, equipment, knowledge, and experience (e.g., Abell, 1990). The primary focus on developing and implementing science lessons would further eliminate pressures to cover other elementary curricular subjects. As such, elementary science specialists might be able to deliver more innovative and effective science instruction than typical elementary classroom teachers can or should be expected to deliver. Schwartz, Abd-El-Khalick, and Lederman (2000) sought to evaluate the effectiveness of using science specialists to teach elementary science. This was a "high-stakes" evaluation, given the school board was seeking to justify the costs of a specialist-led elementary science program in light of the fact that such programs were rarely implemented in other school districts in the state. Each of the specialists studied taught science in their own laboratory classroom facility, similar to the physical teacher education model (Abell, 1990) in which classes visit a science laboratory on a rotating basis, just as classes go to the gymnasium for physical education. The regular classroom teachers in the comparison district simply taught science in their own classrooms.

Schwartz et al. (2000) found specialists' views of elementary science instruction were more aligned with the reforms vision for elementary science instruction than those of the regular classroom teachers, and concluded the apparent absence of constraints to teaching science (adequate planning time, materials, space, etc.) voiced by classroom teachers suggests specialists could play a significant role in achieving reform. Though no significant differences were found between standardized test scores of students in the specialists' classrooms and those in which science was taught by classroom teachers, the authors caution that these tests were measures of lower-level knowledge and comprehension. Analysis of student work artifacts from students in specialists' classrooms illustrates learning outcomes related to higher-order skills involved in conducting scientific inquiry; however, similar artifacts were not provided by the comparison district.

In their discussion, the researchers acknowledged the cost of hiring specialists might serve as a barrier for some school districts. It is not clear whether the cost of the specialists or the cost of the facilities was more of an issue, however, because the researchers did not examine these aspects separately. In addition, it is difficult to draw conclusions from this study about how effectively laboratory facilities would be utilized without the specialists or vice versa. For example, some delivery models for science may involve a specialist who visits students' classrooms, rather than teaching science in his or her own laboratory facility (Gess-Newsome, 1999).

Jones and Edmunds (2006) explored three different delivery models for elementary science instruction, two of which included a separate science laboratory facility. In the "resource model," the science lab housed teacher and student resource books, as well as manipulatives, kits, and consumable materials, all of which were available for classroom teachers' use, similar to the lab facility studied by Vorsino (1992). As Jones and Edmonds report, "the teachers and the principal feel that the lab provides the necessary resources for science instruction to happen in the classroom" (p. 325). They found this model facilitated interdisciplinary instruction and fostered collaboration between the science resource teacher and classroom teachers within the school. In contrast, the "science instructor model" involved a specialist who taught in her own science laboratory (similar to Schwartz et al., 2000). Though classroom teachers were responsible for teaching science, there were concerns that some teachers might avoid doing so because of the presence of the science specialist. Despite these differences, the authors argued that both of these models resulted in a more significant "presence for science" within the two buildings, particularly in comparison to a more traditional "classroom teacher model" of elementary science instruction. In the schools with special laboratory facilities there was:

> more leaderships for science, more community involvement in the science program, and teachers in the schools with specialists tended to involve students with special interests and aptitudes in science, worked with colleagues across grade levels to plan the science program, and had teachers more involved in professional development in science. (Jones & Edmunds, 2006, pp. 332-333)

A NEED FOR RESEARCH

Currently, the science education literature yields no clear insight about the way in which science laboratory facilities in elementary schools contribute to ensuring adequate and appropriate laboratory instruction at the elementary level. The paucity of studies focusing specifically on the use of separate facilities for elementary science limits the abilities of school districts to make evidence-based decisions about how to invest financial, personnel, and building resources. Research is needed to further examine the use and implementation of science laboratory facilities for elementary schools and the way in which these support the provision of laboratory experiences for students. Several possible venues for study are suggested below.

Barriers to Effective Science Teaching

According to the *National Science Education Standards* "effective science teaching depends on the availability and organization of materials, equipment, media, and technology" (NRC, 1996, p. 44). The establishment of specialized laboratory facilities at the elementary level is one strategy for pooling together and facilitating sharing of materials and resources for the benefit of all teachers and students. Research by Vorsino (1992) suggests that while elementary laboratory facilities may be designed to address barriers to effective science teaching, implementing laboratories may in fact create a new set of barriers as well. Thus, future studies may explore:

- What barriers to effective laboratory instruction are alleviated by implementation of science laboratory facilities at the elementary level?
- What new barriers are encountered in utilizing a separate laboratory facility for science instruction?
- What implementation strategies for laboratory facilities are most effective in overcoming barriers to effective science teaching?

In particular, given the findings of Vorsino (1992) it will be important to consider the types of professional development needed to support teachers in effective utilization of the laboratory, and how these assist teachers in developing strategies for conducting laboratory investigations with their students, as well as the types of collaboration among teachers that results from such school-wide initiatives. For example, a shared or common science laboratory might transform the norms of privacy and isolation that typically characterize teaching within one's own classroom.

The Nature of Teaching in the Laboratory

It stands to reason that the very concept of a "laboratory" might evoke a particular image of science, and thus a particular set of norms and expectations for the types of scientific work conducted within the laboratory. According to Gough (1992),

> most school laboratories are crude stereotypes of the diverse sites in which scientists pursue their labors.... [L]aboratories in schools have become places where scientific work is fictionalized in ways that seem likely to

impede learners from understanding the meaning and significance of scientific production in our society and culture. (p. 2)

Thus, it would be important to consider the way in which the physical space itself shapes the activity within it. In considering the nature of science instruction conducted in the lab, it would be fruitful to consider:

- How does science instruction in the laboratory differ from science instruction carried out in other settings, such as a multipurpose classroom?
- How does the nature of the science laboratory facility shape the nature of science teaching?

The Nature of Learning in the Laboratory

The *National Science Education Standards* also emphasize "the arrangement of available space and furnishing in the classroom or laboratory influences the nature of the learning that takes place" (NRC, 1996, p. 44). Thus, it is important to understand the ways in which separate facilities enhance and/or enable science-learning experiences that would otherwise not be possible in the regular classroom. Possible questions might include:

- What types of learning experiences are made possible by the use of specialized facilities at the elementary level?
- How do these laboratory learning experiences differ from those which would be possible in a multipurpose elementary classroom?

Given elementary students are likely to hold stereotypical views of science and scientists, such as scientific work taking place in a laboratory (Barman, 1997), it may be important to understand how receiving instruction in a laboratory setting impacts students' perceptions of science, in addition to their learning of science. For example,

- What are the norms for learning science in the laboratory, and how do these compare to the norms for learning other subjects in the classroom?
- In what ways does the use of separate facilities for science influences students' views of scientific work? The nature of science?
- How does the use of separate facilities for science influence students' attitudes toward science as a discipline? As a human endeavor?

The Role of Elementary Science Laboratories in the Science Program

It should be emphasized that facilities for science instruction are utilized in the context of a delivery model for science instruction within a school. In the United States, approximately 15% of elementary students receive instruction from a science specialist in addition to their regular teacher, and another 12% receive science instruction from a science specialist instead of their regular classroom teacher (Weiss et al., 2001). Prior research illustrates the importance of considering "effective" use of an elementary science laboratory in relation to the delivery model for instruction (Jones & Edmonds, 2006; Schwartz et al., 2000).

- In what ways does the use of the science laboratory facility support the delivery model for science within the school?
- In what ways does the existence of the science laboratory facility contribute to a broader presence for science at the school?

Clearly this is not strictly a pedagogical issue. Determining the cost-effectiveness of separate laboratory facilities as opposed to better equipping elementary classrooms to be more suitable for science instruction is of importance not only in the construction of new schools, but in the remodeling and refurbishing of existing school facilities as well. Therefore research should also consider:

- In what ways does the science lab facilitate the flow of resources, financial and otherwise, to the school science program?
- In what ways does the use of separate laboratory facilities impact student access to science?
- What considerations should be made when deciding whether to designate a room in an elementary school as a science lab?

CONCLUSIONS

Previous research has identified a number of barriers to effective science instruction, particularly at the elementary level. The level of adequacy of facilities and sufficiency of materials and resources for teaching science are important limiting factors in the successful implementation of laboratory investigations. Though it is clear that elementary schools are investing resources into the creation of separate facilities for science instruction, little is known about the impact of these facilities and their implementation on the quality of science teaching and learning. Understanding the role of laboratory activities in science learning can be enhanced by better understanding the context in which that learning takes place. Similarly, understanding how to better achieve effective implementation of laboratory investigations can result from research that provides insights into the way that the classroom (or laboratory) setting affects the enactment of science teaching and learning. Comparative studies of elementary schools that do and do not utilize separate laboratory facilities in their delivery models may shed light on these issues, as well as case studies of individual schools that have transitioned science instruction from regular classrooms to a laboratory system or vice versa.

ACKNOWLEDGMENTS

Portions of this chapter are based on a manuscript forthcoming in the *Journal of Elementary Science Education*, written by the author and with permission of the journal.

REFERENCES

Abell, S. K. (1990). A case for the elementary science specialist. *School Science and Mathematics, 90*, 291-301.

Banilower, E. R., Smith, P. S., & Weiss, I. R. (2002). *Examining the influence of National Standards: Data from the 2000 National Survey of Mathematics and Science Education.* Chapel Hill, NC: Horizon Research.

Barman, C. (1997). Students' views of scientists and science. *Science and Children,* 18-23.

Biehle, J. T., Motz, L. L., & West, S. S. (1999). *NSTA Guide to School Science Facilities.* Arlington, VA: National Science Teachers Association.

Czerniak, C., & Chiarelott, L. (1990). Teacher education for effective science instruction: A social cognitive perspective. *Journal of Teacher Education, 41*, 49-58.

Fehlig, J. C. (1996). Parents' science lab: Invite parents to lead hands-on science activities at school and watch students' interest grow. *Science and Children, 34*(2), 17-19.

Fox, P. (1994). Creating a laboratory: It's elementary. *Science and Children 31*(4), 20-22.

Fulp, S. (2002). *The status of elementary science teaching.* Chapel Hill, NC: Horizon Research.

Gess-Newsome, J. (1999) Delivery models for elementary science instruction: A call for research. *Electronic Journal of Science Education, 3*(3). Retrieved June 6, 2007, from http://wolfweb.unr.edu/homepage/crowther/ejse/newsome.html

Gough, N. (1992, April). *Laboratories in schools: Material places, mystical spaces.* Paper presented at the annual meeting of the American Educational Research Association. San Francisco, CA.

Harbeck, M. B. (1985). Getting the most out of elementary science. *Science and Children, 23*(2), 44-45.

Higle, S. (1984). In the schools: An elementary science lab in action. *Science and Children, 22*(2), 44-45.

Hofstein, A., & Lunetta, V. N. (2004). The laboratory in science education: Foundations for the twenty-first century. *Science Education, 88,* 28-54.

Jones, M. G., & Edmunds, J. (2006). Models of elementary science instruction: Roles of science specialist teachers. In K. Appleton (Ed.), *Elementary science teacher education: International perspectives on contemporary issues and practice* (pp. 317-344). Mahwah, NJ: Erlbaum.

Mackin, J., & Williams, F. (1995). Science in any classroom: How to cope with less than adequate facilities. *The Science Teacher, 62*(9), 44-46.

National Research Council. (1996). *National Science Education Standards.* Washington, DC: National Academy Press.

National Research Council. (2006). *America's lab report: Investigations in high school science.* Washington, DC: National Academy Press.

National Science Teachers Association. (February 2007). *NSTA position statement: The integral role of laboratory investigations in science instruction.* Arlington, VA: Author.

Richardson, J. S. (Ed.). (1951). *School facilities for science instruction.* Washington, DC: National Science Teachers Association.

Schwartz, R. S., Abd-El-Khalick, F., & Lederman, N. G. (2000). Achieving the reforms vision: The effectiveness of a specialists-led elementary science program. *School Science and Mathematics, 100*(4), 181-193.

Smith, P. S., Banilower, E. R., McMahon, K. C., & Weiss, I. R. (2002). *The National Survey of Science and Mathematics Education: Trends from 1977 to 2000.* Chapel Hill, NC: Horizon Research.

Tilgner, P. J. (1990). Avoiding science in the elementary school. *Science Education, 74*(4), 421-431.

Tobin, K. G. (1990). Research on science laboratory activities. In pursuit of better questions and answers to improve learning. *School Science and Mathematics, 90,* 403–418.

Vorsino, W. S. (1992). *Improving the effectiveness of science laboratory instruction for elementary students through the use of a process approach for change.* Unpublished thesis, Nova University, Fort Lauderdale-Davie, FL.

Weiss, I. R., Banilower, E. R., McMahon, K. C., & Smith, P. S. (2001). *Report of the 2000 National Survey of Science and Mathematics Education.* Chapel Hill, NC: Horizon Research, Inc.

PART II

THE STATUS AND IMPACT OF LABORATORY WORK IN LEARNING AND TEACHING SCIENCE

INTRODUCTION

While the first section of this volume introduced the reader to the importance of the design of the laboratory and physical context for effective science learning, the second section of this volume centers on the current status of research on the impact of the laboratory on student learning. Overall, the authors in this section describe impact on student learning utilizing nontraditional pedagogy, primarily from a constructivist perspective. The section begins in chapter 4 with a description by Billie Eilam of the impact of long-term laboratory inquiry on student understanding of ecology. Eilam advocates the use of long-term laboratory inquiry to address barriers to student learning resulting from the typical fractionated nature of short-term, individual laboratory experiences. In chapter 5, Sule Ozkan, Jale Cakiroglu, and Ceren Tekkaya describe the impact of students' perceptions of the laboratory-learning environment on student learning. In chapter 6, Andrea Gay reports on the impact of a constructivist write-to-learn technique using student construction of meaning in organic chemistry. The final chapter in this section written by Tarek Daoud and Saouma BouJaoude examine the effect of using the Vee Heuristic, a visual graphic organizing learning tool, on students' meaningful learning in physics laboratories.

In chapter 4, Eilam describes the impact of long-term laboratory inquiry on student understanding of ecology. The introduction to the chapter examines effective strategies in teaching ecology and typical barriers to effective ecology instruction. Next, Eilam continues with a

The Impact of the Laboratory and Technology on Learning and Teaching Science K-16, pp. 73–75

description of the model of long-term laboratory inquiry examined in the research and the results on student learning. Eilam concludes with a summary of the findings of the impact of long-term inquiry on student understanding of ecology and makes connections with the results and conclusions of other published studies.

In Chapter 5, Ozkan, Cakiroglu, and Tekkaya report findings from research into the impact of students' own perceptions of the science laboratory-learning environment and science performance outcomes. Also, the authors noted gender was a significant factor in student perceptions of the laboratory-learning environment. The chapter begins with a summary of the research on students' perceptions of laboratory experiences in several countries, the United States, Taiwan, and Korea, including the research described by Ozkan, Cakiroglu, and Tekkaya conducted in Turkey. The instruments used in the study, Science Laboratory Environment Inventory and the Science Attitude Scale were translated into Turkish and subsequently validated for use with Turkish students. In general, the results mirrored research conducted in other countries; data analysis did reveal gender differences. The chapter concludes with recommendations for further research into designing laboratory experiences that are effective for both genders.

Gay addresses writing in chemistry laboratories in chapter 6 which begins with a summary of research in writing-to-learn methodologies. Gay reported three broad categories were involved in writing-to-learn methodologies. First, in the modernist approach, students read and write in the typical technical language of science found in traditional laboratory reports and journal articles. Second, students use writing to translate science concepts in their own words as a means of constructing meaning in the constructivist approach. The last writing-to-learn methodology, poststructuralist methodology, is primarily concerned with student understanding of the impact of science on society and is rarely used in traditional laboratory writing. Following a description of writing-to learn methodologies, Gay discussed the development of the decision/explanation/observation/inference laboratory writing methodology (DEOI) used in the research. This research design investigated the impact of the DEOI constructivist approach to laboratory journal writing involved in student understanding of the connections between chemical theory and laboratory application. The case study approach described the differences in student understanding between classes taught by two graduate teaching assistants. Gay concluded that student understanding of connections between chemical theory and laboratory experiences was greater when the teaching assistant was more effective in implementing the writing-to-learn DEOI methodology. Finally, Gay noted, however, students in both sections reported the DEOI methodology assisted in constructing meaning.

In chapter 7, Daoud and BouJaoude examined the impact of the Vee Heuristic on student learning in physics. The chapter begins with an overview of the Vee Heuristic and a review related research. Daoud and BouJaoude described their research in the use of the Vee Heuristic in a Grade 10 physics laboratory in terms of student learning and student attitudes toward physics. Results indicated no statistical difference was found between the experimental group and control group in terms of student achievement when the Vee Heuristic was used. In terms of attitude however, student attitude towards physics remained the same in the experimental group using the Vee Heuristic, but student attitudes dropped in the control group. Thus, Daoud and BouJaoude concluded utilization of the Vee Heuristic positively impacted student attitudes towards physics. The chapter concludes with recommendations for further research in the use of Vee Heuristic in science classrooms and in particular, Daoud and BouJaoude reported specific training was required in order for students to effectively utilize the Vee Heuristic.

Part II begins with a case study on the impact of long-term laboratory experiences on student understanding of ecology in contrast with traditional, fractional, individual laboratory activities on specific concepts. Eilam advocated the use of long-term laboratory experiences to aid student construction of meaning through links between ecology constructs as a result of extended laboratory experiences on specific topics. In a second case study of Turkish students, Ozkan, Carikoglu, and Tekkaya describe the impact of students' perceptions of the laboratory-learning environment on student learning; results reveal similar trends in Turkey previously reported by researchers in other countries. The impact of the process of writing in the laboratory on student understanding was the focus of the third case study presented by Gay in chapter 6. Gay concludes a positive relationship exists between student construction of meaning and journaling in organic chemistry. In the final chapter of this section, Daoud and BouJaoude examine the positive impact of the Vee Heuristic on student attitudes in a Grade 10 physics class and note a neutral effect on student learning. In summary, the chapters in section two describe non-traditional science laboratory pedagogical strategies utilized to positively impact either student understanding or student attitudes in science and provide readers with suggestions for further research in these areas.

CHAPTER 4

LONG-TERM
LABORATORY INQUIRY

Promoting Understanding of Ecology

Billie Eilam

The chapter describes a pedagogy consisting of a long-term inquiry laboratory (LTIL), specifically designed to enhance the understanding of ecosystems. In addition, the design enhanced students' acquisition of inquiry and
self-regulating skills. The study focuses on the construct of feeding relations
in ecosystems. A detailed description of the LTIL pedagogy and of empirical studies that examined various aspects of it is presented, with findings
regarding LTIL successful application. Suggestions, regarding a possible
instruction of ecology that accounts for the reported learners' difficulties
and poor performance concerning system thinking, are presented.

INTRODUCTION

Science educators are united concerning the importance of learning ecology, recognizing the scientific endeavor and the impact of social, political,
economic, and moral contexts on ecology. Students in the postmodern

*The Impact of the Laboratory and Technology on Learning and
Teaching Science K-16*, pp. 77–109
Copyright © 2008 by Information Age Publishing
77

world are required to participate intelligently in decision-making processes pertaining to their own quality of life and to environmental issues (Driver, Leach, Millar, & Scott, 1996). However, numerous difficulties inherent to the study of ecology are often reported. Many have suggested that some of these difficulties may be overcome by learning ecology via inquiry, namely, by the acquisition of an understanding of both the empirical procedures of inquiry and the theoretical and conceptual ideas at its foundation (Bentley & Watts, 1992).

The present chapter describes a pedagogy consisting of a long-term inquiry laboratory (LTIL), specifically designed to enhance the learning of ecology as well as the acquisition of inquiry skills, and present findings regarding its successful application. The chapter begins with a presentation of reports concerning the difficulties in ecology study and examples of prior implementation of relevant laboratories and inquiries. Next, the chapter continues with a detailed description of the LTIL pedagogy and of empirical studies that examined various aspects of the strategies used. The chapter ends with a discussion of the findings of the studies summarizing LTIL's ability to promote ecology study and skill acquisition.

LEARNERS' DIFFICULTIES IN ECOLOGY AND IN INSTRUCTIONAL-RELATED LABORATORIES AND FIELD WORK

The difficulties inherent to the learning of ecology are reflected in students' poor performance (Esiobu & Soyibo, 1995). Some of students' difficulties and alternative frameworks have been attributed to the simplistic representations of science and its processes in textbooks and classroom discourses (McComas, 1998). In addition, ecological knowledge is difficult to construct because ecological theories cannot be easily applied, having little formalization and systematization. Changes in the more general ecological statements result in ecologists having to constantly revise their understanding of ecosystems (Pickett, Kolasa, & Jones, 1994). Therefore, Gonzalez del Solar and Marone (2001) warned against the dogmatic teaching of ecology, which may result in solidified bodies of knowledge that, when new theories replace old ones, become less efficient in explaining new situations. To improve students' understanding, these researchers call for instructors to use pedagogy that parallels scientific research, emphasizing theory.

Many studies have demonstrated that students' naive ideas and underlying beliefs regarding basic ecological concepts like food chains/webs seem to persevere despite specially designed instruction (Anderson, Sheldon, & Dubay, 1990; Gallegos, Jerezano, & Flores, 1994; Griffith & Grant, 1985; Hogan, 2000; Hogan & Fisherkeller, 1996; Leach, Driver, Scott, &

Wood-Robinson, 1996; Munson, 1994). For example, students' ideas concerning respiration were reasoned by their inability to link new information about respiration to concepts related to digestion, photosynthesis, and energy in food chains, their use of everyday language (breathing), or their confused utilization of many different terms for the same concept (internal respiration, external respiration, cellular respiration, general respiration, aerobic respiration, or just respiration) (Alparslan, Tekkaya, & Geban, 2003).

The systemic nature of ecosystems, and the transfer of matter and energy within them, has gained special consideration as the essence of ecology. Robinson (as cited in Driver & Millar, 1986) asserted that although biology students encounter ideas on energy in many biology-related contexts (e.g., photosynthesis, nutrition), which are essential for understanding ecology and food chains, energy is portrayed differently than in physics classes, where it is viewed in terms of the laws of thermodynamics and is treated quantitatively, thus hindering comprehension. In comparison, in biology, there is usually less emphasize on energy being substantially different from matter and it is mostly mentioned along with the description of various processes (e.g., photosynthesis), resulting in students' deficient understanding of its properties and their meaning for biological processes.

Although the understanding of evolution is a prerequisite for the deep comprehension of ecology, it is very problematic as presented in ecology. Students' learning difficulties and the dominancy of the Lamarckian view are frequently reported by students in ecology classrooms (Demastes, Good, & Peebles, 1996). These difficulties may rise from the evolution's characteristics, namely, its counterintuitive and abstract nature, lengthy time scale, and lack of evidence, from the inadequate transmission of information about evolution (Crawford, Zembal-Saul, Munford, & Friedrichsen, 2005), or from students' beliefs and preconceptions concerning religion and the nature of science (Alters, 1999; Blackwell, Powell, & Dukes, 2003). Students' comprehension did not improve significantly even after demonstrations of natural selection and short-term evolutionary changes in long-term field studies, computer models, and case simulations (Bishop & Anderson, 1990; Passmore & Stewart, 2002; Unger, Seligman, & Noy-Meir, 2004).

Ecology is interdisciplinary in essence, thus requiring an interdisciplinary approach (Klein, 1990). Concepts like energy, regarding ecosystems, interact on the border zone of several disciplines (e.g., biology, physics, chemistry, geography); hence, learners must draw on these disciplines to develop deep understanding of these concepts.

In terms of these many difficulties, traditional ecology lessons accompanied by some hands-on labs have not seemed to provide an adequate solution. Although some of students' naive concepts were found to change slightly, their deep commitment to underlying ontological beliefs did not, thus allowing a naïve conceptual environment to dominate classroom talk (Reiner & Eilam, 2001). To address these problems, some researchers and educators have advocated student laboratory experimentation in ecology, notwithstanding its complex logistics and ethical constraints, as reflecting the more adequate domain practice. Several principles have been suggested to guide studies of ecology lab practices (Finn, Maxwell, & Calver, 2002): employing treatment and control groups in studies to distinguish experiments from sampling; following a logical pattern of deductive reasoning despite the domain's complexity; representing the ecologist's tools; and relating ecology to environmental issues. Others have suggested that to attain complex concepts students must integrate lab studies with the acquisition of theoretical knowledge, regardless of the physical hardships in organizing a school setting that would accommodate for this need (Cottrell, 2004; Fail, 2003).

The application of problem-based learning or scientific inquiry as an advantageous core pedagogy has been advocated for promoting scientific skills and metacognition (Minstrell & van Zee, 2000; Uyeda, Madden, Brigham, Luft, & Washburne, 2002). However, the high cognitive demands that such a pedagogy places on students raise the need for substantial support regarding the various elements of such a practice and in particular the need for collaboration with others in order to reduce cognitive load, support for students' application of self-regulation and other metacognitive abilities, as well as of the inquiry skills (Krajcik, et al., 1998). The many difficulties inherent to such problem-based learning frequently result in teachers' assignment of only small-scale, single-topic inquiries that usually do not relate to the larger context. For example, the separate study of plant needs, nutrition, and photosynthesis, removed from matter and energy cycles in the biosphere, does not promote ecology understanding (Lee, 2003; Ozay & Oztas, 2003). On the other hand, ecological inquiry focused on a specific, authentic issue can increase students' motivation and teach ecology principles and techniques. For example, in a task requiring students to "sell" a portion of a forested park to developers (Tessier, 2004), students had to develop tools for assessing the ecological value of property, assess the park area, and decide which portion to sell with minimal repercussions, using sampling techniques.

To address the aforementioned concerns, researchers have usually suggested learning ecology through an open-ended inquiry. Roth (1995) reported on the collaborative learning of eighth graders in a 10-week, open-ended, complex ecological inquiry. Such inquiry constituted a challenge to students in that problems and solutions could not simply be predicted by students' prior knowledge and experiences. Students investigated the interrelations among biotic and abiotic composites of a defined field-site ("ecozone") both in the field and in the classroom. Roth conceptualized classroom environment as a culture with complex interactions among its individuals that were afforded certain practices by the available classroom environmental resources. Some of these practices were shared between students to become part of their community culture, and new resources were constantly recognized by and became familiar to community members in the course of their inquiries, making new practices available for investigating a phenomenon. A microanalysis of conversations (Roth, 1995) demonstrated students' serious involvement in the learning processes and their deep understanding of the phenomena.

THE PEDAGOGY OF AN OPEN-ENDED, LONG-TERM INQUIRY ECOLOGICAL LABORATORY

The previously described difficulties call for the design of a special learning environment to promote students' understanding of the complex domain of ecology. As a representative of ecosystems' structure and function, the construct of feeding relations was chosen. Feeding relations involve the interactions among biotic and abiotic components, transfer of energy and matter, effects of mutations and selective forces, and the simultaneous multidirectional occurrences of systemic processes. The new pedagogy described herein utilized a laboratory inquiry in line with the constructivist approach; and was based on research findings regarding cognition and learning processes conducive to the acquisition of "deep understanding." According to constructivism, a lab inquiry engages students actively in processes of constructing knowledge, integrating it with existing knowledge, and applying knowledge; such processes bring about gradual knowledge revisions and conceptual changes (Brooks & Brooks, 1993; Driver, Asoko, Leach, Mortimer, & Scott, 1994). "Deep understanding," based on Perkins' (1998) definition, may be recognized through performance, referring to students' ability to think and act flexibly concerning numerous simultaneous and consecutive aspects of an ecosystem, apply these aspects as integrated knowledge, and relate them to the various bodies of the involved disciplinary knowledge.

This new pedagogy (LTIL) was carried out in a junior high school in northern Israel for several consecutive years, accompanied by a comparison to the traditional learning mode and by investigation of major issues related to the deep comprehension of ecological issues and to skill acquisition and application. In a certain year, 112 ninth grade students were

randomly divided into an experimental condition where students experienced a yearlong LTIL ($n = 60$) and a yearlong control condition ($n = 52$)
where students learned in the traditional mode.

THE CURRICULUM

The Traditional Learning Mode

Students ($n = 52$) attended six traditional science lessons each week
throughout the academic year: (a) three in ecology, mostly teacher controlled with regard to all the aspects of learning and environment; and,
(b) three in physics, teacher-controlled as well. In each subject, students
experienced some hands-on laboratory assignments that focused on single concepts and targeted certain expected results (e.g., effects of light on
photosynthesis, effects of temperature on plant growth rate).

The New Pedagogy

Students ($n = 60$) attended the same number of lessons each week
throughout the year, as follows: (a) three traditional lessons, one in ecology and two in physics, which granted students the knowledge base necessary for independent study and focused on constructing a systemic view of
ecology, rather than remaining at the level of single factors; (b) three consecutive hours of the independent, open-ended inquiry lab (LTIL), in
which students investigated a subject of their choice concerning ecosystem functioning. The students discussed issues with peers and teachers
within a constructivist approach, raising topics according to their own
understanding, timing, and needs, as stemming from materials covered
in the theoretical lessons or by their own inquiries.

The contents of these theoretical lessons, based on a textbook
developed for teaching ecology through the systemic view (Eilam,
Lilienfeld, Zernik, Carmon, & Ramon, 1980), focused on issues like: the
function of open systems (e.g., homeostasis, feedback mechanisms,
equilibrium), biotic and abiotic components of ecosystems and their
interrelations, energy and matter conservation and transformation, and
so forth. Evolution, plant structure and function, the human body, and
bacteria were studied during the seventh and eighth grades. These
contents were organized to gradually increase complexity to achieve the
systemic view. Single basic concepts were taught first (e.g., respiration,
photosynthesis, matter and energy conservation), and served as a
foundation for identifying components and explicating their

interrelations to promote systemic understanding. Situations were conceptualized as collections of multiple simultaneous occurrences rather than as linear processes (see the Appendix for an example). Prior awareness of students' reported naive knowledge and beliefs concerning concepts (Hatano, Siegler, Richards, Inagaki, Stavy, & Wax, 1993; Hogan & Fisherkeller, 1996; Reiner & Eilam, 2001) was accounted for, thus enabling teachers to carefully weave the instructional contents to promote students' knowledge construction.

In all biology lessons/labs through the year, to achieve students' systemic mode of thinking about ecosystems, theoretical exercises were performed that required the application of complex, simultaneous paths and levels of thought. These systemic exercises aimed to transform the concrete into the abstract and complex, by enhancing mastery of the principles of systemic functioning rather than merely promoting specific knowledge concerning particular incidents (e.g., air or water pollution, cutting down the Amazon forest). Students were required to examine the manifestations of their acquired conceptual knowledge in the analysis of authentic episodes in diverse ecosystems (e.g., monoculture, aquatic, urban) in Israel (e.g., the drying of the marshes in the Hula valley and its effect on the Sea of Galilee, kibbutz monocultures like cotton, or cows in a dairy farm) and from around the world, especially in neighboring countries (e.g., the effect of the Aswan dam's construction on The Nile River and its delta). The focus in these exercises was specifically on interrelations among each ecosystem's various components (e.g., feeding relations, matter transfer). Generative knowledge (Perkins, 1992) was targeted, including understanding of generic rules, function patterns, and relations, in order to advance students' future ability to analyze and deliberate on new and unfamiliar ecological cases. These exercises also aimed to promote students' application of the acquired abstract, complex knowledge to the concrete cases of their own inquiry ecosystems (see below).

LTIL Characteristics

Students in the LTIL conducted an inquiry into an ecosystem (i.e., aquarium, terrarium, or greenhouse) that they selected, constructed, and monitored in small groups, while acquiring inquiry skills and applying them in their own inquiries. Students chose the inquiry subject, formulated research questions and hypotheses based on theoretical information they gathered while planning and constructing their ecosystems, and collected data in these ecosystems by conducting weekly measurements of a range of relevant variables over the course of several months (e.g., temperature, humidity, oxygen concentration in water, amounts of food and water added to the ecosystem, CO_2, phosphates, nitrates). When required to reason about the measured changes in these variables' levels, students became involved in a deep and dynamic process of making sense of the interrelations and complex processes occurring in their ecosystems. Data measurement and processing, was supported by the physics classes and enabled theoretical knowledge to become meaningful, relevant, and reality-based. These included the use of measuring instruments, calculation of ranges of possible errors or the deep application of knowledge concerning energy properties.

Students were prompted to be constantly aware of their systems' input and output (matter and energy), even if not always quantitatively. For example, they were asked to identify the input sources of energy entering their system (e.g., the sun's rays as they shone on various objects; the light bulbs in the ecosystem, room, and aquarium heater; the energy contained in the organic food for the organisms) to follow possible transfer and transformation routes of the identified energy within the ecosystems (e.g., the release of heat from moving fins of fish muscle cells or digesting cells), as well as to identify possible output sources of energy (e.g., heat released) and matter (e.g., solids taken out, gasses diffused, water evaporated). The students were asked to reason about the quantitative and qualitative differences between the input and output of the open ecosystem (e.g., energy tied up in organic matter that builds an organism's body, in mucous excreted through the fish skin or in dry epidermis), while constantly made aware of the relations between matter and energy.

Hence, the LTIL required students to apply theoretical knowledge, which was gained via traditional lessons and exercises, to diverse and constantly changing real life situations in their ecosystems. In turn, the more refined and organized knowledge that developed through these experiences promoted students' ability to acquire and integrate new knowledge more efficiently during the next traditional lesson, and so forth (Eilam, 2002b, 2007b). These reciprocal relations, between the knowledge concerning theoretical systems studied and the students' own ecosystems, generated a gradual accumulation of more systemic generalizations (Eilam, 2002b, 2007b), as suggested also by other studies (Cottrell, 2004; Fail, 2003). Moreover, the analysis of incidents "as they came along" in the ecosystem, rather than as linear teacher-determined ones typifying traditional classrooms, set in motion students' ongoing construction of holistic, systemic representations of occurrences in ecosystems (Eilam, 2002b, 2007b).

The LTIL condition led students to conduct many small-scale experiments targeting limited questions, which, altogether, helped students design their main inquiry experiment. The duration and scope of these small experiments resembled the typical demonstrations and experiments performed in a regular hands-on lab, but differed in several aspects. They were "invented" (designed) by students at a time of readiness to deal with the specific issue and were based on the structure and organization of students' available declarative and procedural knowledge at that particular moment. Hence, students exhibited an increased ability to interpret the results and to integrate the newly produced knowledge with their own existing knowledge, thus enabling students to move forward in performing the larger investigation (Eilam, 2004a).

Resources

Throughout school hours and during some school vacations, students had access to a large variety of equipment and materials that allowed them to design their own ecosystem, build it accordingly, perform their open-ended inquiry, and gather data (e.g., kits and instruments for growing bacteria or for measuring various variables like the amount of O_2, pH, temperature, humidity, nitrates, phosphates, murkiness, colors; Polyester sheets, sand, and wooden poles for building the ecosystem). The LTIL catered to unexpected needs. Diverse information sources were also available, including written texts (e.g., scientific journals, industrial reports, newspapers); video recordings of ecological cases around the world (e.g., toxic wastes, species extinction); and demonstrations of specific phenomena (e.g., colonies of bacteria, fungi under the microscope, the light spectrum). They also had computers linked to the Internet.

Time Allocation

The lengthy 3-hours per week of lab sessions all through the year enabled students to become seriously involved in making informed choices by investigating a range of alternatives for conducting their inquiry task and achieving their goals (Pintrich, Marx, & Boyle, 1993). Students were free to utilize their time any way they saw fit.

Acquiring Scientific Inquiry Skills Through Practice

For the acquisition of the high-order skill of inquiry, students used a specially developed textbook (Eilam & Aharon, 1998) based on literature concerning the structure of procedural knowledge (Anderson, 1993), as well as theories concerning skill instruction and acquisition (Baron & Sternberg, 1987; Perkins, 1987; Perkins & Salomon, 1989; Salomon & Perkins, 1989). Each chapter presents a single stage of inquiry (e.g., designing an experiment, drawing conclusions) and identifies and explicates the specific skills necessary for performing that stage. For example, the four skills needed for data gathering are: (a) choosing the mode of data collection—survey, observations, and so forth; (b) planning data collection methods—designing data sources, selecting tools, and so forth; (c) examining the plans for data gathering in light of the hypothesized inquiry question and conditions; and (d) constructing and designing tools to be used for reporting the collected data. The textbook explicates the cognitive operations (Anderson's production rules) to be carried out while performing each skill. For example, the four cognitive operations comprising the skill of choosing a sample for gathering data are: (a) determining the items that will constitute the sample; (b) designing the sample composition to represent the investigated phenomenon; (c) determining sample size; and (d) determining the number of rehearsals to be carried out in the process of data gathering. Demonstrations for the application of each cognitive operation and of the skill as a whole, in similar and then in different situations from those in which it was acquired, are presented to train students in the procedures through the designed theoretical exercises. An example for a near transfer exercise regarding the skill of selecting the best sample is when students were asked to choose one of the following samples for a study on why red haired individuals have more freckles, and then justify their choice: (a) males with red hair, (b) people with red hair who have no freckles, (c) people who have various amounts of freckles, or (d) people with red hair.

To summarize, students proceeded in two parallel routes during each LTIL session: acquiring the skill and training in its application, then applying it practically in the context of their own inquiry, thereby advancing it. This resulted in the performance of a serious scientific inquiry while constructing an appreciation of the scientific method (Eilam, 2004b).

Small-Group Learning Setting

The setting of small groups of three to five students was chosen (a) to enable students the social construction of knowledge through negotiations (Vygotsky, 1962), and (b) to decrease the high cognitive load placed on individual students in the complex LTIL.

Self-Regulated Learning (SRL) and Support Mechanisms

The extensive duration of the LTIL and students' inexperience in such independent learning necessitated the application of SRL, including strategies for planning and time management. The latter have been found to be important cognitive regulatory aspects of SRL leading to higher academic achievements (Zimmerman & Risemberg, 1997).

Therefore, based on the literature, two tools were developed to promote students' SRL, metacognition, and monitoring of their own progress: the yearly report and the daily report (Eilam, 2007a; Eilam & Aharon, 2003). Both report types were designed to elicit students' internal cues for monitoring their progress toward their daily goal as well as the yearly one. These cues were easily elicited as students recognized the gaps presented in the reports, by comparing the suggested theoretical plan of action over the yearly or daily time scales versus their actual enacted actions (i.e., comparing the setting in which an activity was performed, strategy/skills employed, order of activity enactment, and activity duration) (for details, see Eilam & Aharon, 2003).

The daily report enabled students to break down the long-term aim (i.e., completing the inquiry and reporting it) into short-term, manageable, and achievable weekly goals. The yearly report increased students' awareness to their progress toward achieving their final long-term inquiry aim and thus to regulate it. Students began each lab by completing the plan part of the daily report, by stating their goals and designing a plan to achieve them. During the session, after completing each of the performed activities, students wrote the details concerning that activity in the designated places on the report. At the end of the session, they completed the yearly report by writing their actual position on the yearly time scale, as compared to the suggested position for that date. By eliciting internal cues, these reports' design raised students' awareness (Eilam & Aharon, 2003) to the revealed gaps and differences between the suggested and enacted plans and promoted their metacognitive ability to improve their SRL. In time, some measures of self-efficacy were added to the reports (e.g., asking students to indicate on a Likert scale their level of confidence in their ability to perform the plan they designed), which increased students' ability to realistically evaluate their planning and their abilities based on their weekly experiences over time (Eilam & Aharon, 2003). In addition, the requirement that the students reason about the revealed gaps contributed to their ability to relate causes to outcomes.

Thus, in the long-term, these tools advanced students' SRL abilities (Eilam & Aharon, 2003) by enhancing awareness of its necessary elements, training students in their application, and allowing the construction and revision of relevant knowledge about their ability to learn in a specific context. An improvement in SRL was found to improve academic achievement (Zimmerman & Risemberg, 1997), therefore, granting students the opportunity to experience and develop this important skill is highly recommended. Such experiences affect students' future engagements with similar tasks (Butler & Winne, 1995; Corno, 1993; Winne, 1995; Zimmerman, 1989). Winne and Perry (2000) asserted that too little has been achieved in measuring SRL as an event and in characterizing the temporal unfolding patterns of engagement in terms of tactics and strategies that constitute it. LTIL enabled such SRL measurement and its investigation as a process rather than as a point-in-time feature.

Motivation

Ecology usually raises great initial interest in students, due to its compelling relevance and familiarity from daily media. Nevertheless, the learning difficulties quickly transform it into "just another regular school subject." Students' motivation was enhanced by granting them the freedom to choose their own inquiry subject and by transferring to them full responsibility and complete control over their own learning processes and time (e.g., pace, homework assignments, recess time) (Eilam & Aharon, 2003).

Interdisciplinarity and Teacher Support

Interdisciplinarity was implemented by having both the biology and the physics teachers present in the lab sessions to serve as mediators and facilitators; provide learners with immediate, expert, direct guidance; and foster students' cognitive flexibility by promoting their ability to compare the knowledge components acquired and to transfer it between the two disciplines, putting into practice Klein's (1990) view. For example, in physics, students studied the concept of systems using the example of air conditioners or cars and these systems' states: stable, unstable, or indifferent equilibrium. Biology classes emphasized the uniqueness of biological systems with respect to their biotic components, feedback loops that maintain equilibrium, homeostasis, and so on. The linkage between these two views was achieved when students had to make sense of their own ecosystems in the LTIL. In these cases, opportunities were created for student-teacher consultation and in-group discussion, which exposed students to multiple points of view relating explicitly between distinct pieces of knowledge to construct a deeper, multi-perspective understanding.

The described project underwent evaluation that led to some modifications over its years of implementation (e.g., improvements in the SRL tools' design, a shift from teacher-determined small groups to self-chosen groups). The next section will present findings of studies that investigated the ability of the unique LTIL design to promote students' science learning.

RESEARCH FINDINGS: ASPECTS OF LEARNING IN LTIL

Several issues concerning learning ecology via a LTIL have been investigated via a series of research studies:

Study Type I.   Students' deep comprehension of various dimensions of ecology was studied through the topic of feeding relations, as representing and involving core ideas in ecology, including the development of students' explanatory frameworks and the advantages of the designed pedagogy in comparison to the traditional one.

Study Type II.   Students' possible gains from the performance of a small-scale experiment within the LTIL framework were examined.

Study Type III.   Long-term skill acquisition was investigated regarding both low-order skills (i.e., choosing wisely) and high-order skills (i.e., SRL).

Data Collection and Analysis

Data concerning the various aspects of learning were collected via several tools, in accordance with the specific study goals. The tools included: (a) pretests and posttests designed to probe students' understanding, expose their knowledge representations, and examine their employment of this knowledge while involved in processes of complex reasoning or its application in diverse, new situations; (b) a reasoning task about a small-scale experiment on states of matter; (c) long-term video recordings of two small groups, focusing on students' conduct and discourse; (d) students' daily and yearly reports as reflecting their SRL abilities.

The type of analyses applied in each case matched the specific study objectives and research questions, as well as the data characteristics. Content analysis, reports, and discourse analysis were always performed using both predetermined criteria as well as criteria that were elicited from the students' responses. Quantitative analyses were carried out when appropriate (e.g., comparing learning in the experimental and control conditions or determining students' progress after experiencing LTIL).

I. Deep Comprehension of Ecology

Three main studies investigated students' deep understanding of ecology: (a) the difficulties in teaching ecology in a traditional classroom. In this study, ontological and content beliefs of ninth grade students ($N = 28$), as elicited by tests administered before and after learning ecology in a traditional classroom, were analyzed and compared. Results showed that although students' concepts were changed, underlying ontological beliefs are not necessarily changed, and classroom conceptual environment changed only slightly. (b) the impact of LTIL on students' understanding of core ideas in ecology as compared with the impact of the traditional learning mode. This study examined and compared between ninth grade students' (N = 112) deep understanding of feeding relations, after learning in a traditional versus an experimental environment, similar to the one described in the present chapter. Results suggested a significant improvement in understanding of system complexity of students learning in the experimental environment in comparison to those learning in the traditional one, with yet some persistence of specific difficulties related to energy and evolution. (c) students' explanatory frameworks of the construct of feeding relations. Based on reported results concerning students' understanding of the complexity of ecological system, a set of ten explanatory frameworks of feeding relations were suggested as describing students' knowledge development concerning this construct. (For details on these studies, see Eilam, 2002b, 2007b; Reiner & Eilam, 2001. Here, the studies' findings will be discussed together.)

Generally, students' understanding of ecology after learning in the context of a traditional classroom environment, although it included demonstrations and hands-on lab experiments, was found to be deficient regarding the domain's study objectives. Findings revealed that although students' understanding of certain elements in ecosystems improved after learning, on the whole, the traditional classroom's conceptual environment remained the same, exposing individuals to a discourse that may have hindered their advancement (Reiner & Eilam, 2001). In contrast, the rich ingredients of the LTIL pedagogy appeared to improve comprehension and promote processes of conceptual change concerning many concepts of the construct of feeding relations, significantly more than did the traditional learning context (as seen in Table 4.1).

The analysis of students' difficulties in studying ecology in the traditional mode revealed four major dimensions that were suggested as constituting a basis for developing system thinking and an understanding of feeding relations in ecology (Eilam, 2002b): (a) the *macrolevel* dimension—the elements comprising food chains/webs, and their order and evolutionary-determined characteristics, excluding abiotic elements of matter and solar energy; (b) the *microlevel* dimension—the molecular level of processes within organisms and ecosystems (e.g., matter transformation and molecules flowing in and through organisms as part of matter cycles); (c) the *spatial* dimension—the simultaneous occurrence of processes in space, dynamic equilibrium, and organisms' ability to hold several roles at any point in time; and (d) the *temporal* dimension—evolution-related issues (e.g., the selection of more stable food webs, organisms' feeding on specific foods according to their anatomy and physiology, changes in populations in the course of evolution because of changes in genes frequencies due to random mutations and selection forces).

Students' mental models of feeding relations and their deep understanding were revealed using 12 criteria (see Table 4.1) that emerged from these four dimensions and were used for analyzing students' responses to various probes (Eilam, 2007b).

The findings suggested that the LTIL allowed students to engage in a long-term process of knowledge construction and revision while progressing in their ecosystem inquiries, by managing the process to fit their own pace, considerations, and preferred strategies. In the traditional mode, students mostly had to work on their own to link the various bits of knowledge they acquired or constructed during lessons or labs, whereas long-term engagement with the same subject of inquiry enabled individuals in the LTIL to build new knowledge, integrate it within existing knowledge, and accommodate the latter accordingly.

Table 4.1. Criteria and Subcriteria for Content Analysis of Students' Learning Discourses and Posttest Responses as well as a Comparison of Experimental and Control Study Groups

	Criteria	Subcriteria	Posttest Comparison of Experimental (n = 50) and Control (n = 52) Groups		
				M	SD	t
1	Configuration	Linear	Control	3.56	2.79	3.10***
			Experi.	5.08	2.14	
		Cyclic	Control	.60	1.44	1.28*
			Experi.	.28	1.01	
2	Structure	Linear	Control	2.52	2.91	.41
			Experi.	2.28	2.94	
		Branched	Control	.79	1.17	4.22***
			Experi.	.08	.27	
		Web	Control	.04	.28	5.22***
			Experi.	1.86	2.45	
3	Components	Biotic and abiotic	Control	.48	1.29	.47
			Experi.	.36	1.27	
		Only biotic	Control	4.10	2.64	1.51*
			Experi.	4.84	2.35	
4	Component type	Plants	Control	4.17	2.67	2.80***
			Experi.	5.40	1.65	
		Animals	Control	5.31	1.78	.58
			Experi.	5.50	1.52	
		Humans	Control	.56	1.61	4.75***
			Experi.	2.82	2.94	
		Decomposers	Control	3.29	2.91	.82
			Experi.	3.75	2.90	

Table continues on next page.

Table 4.1.   Continued

Criteria	Subcriteria		Posttest Comparison of Experimental (n = 50) and Control (n = 52) Groups		
			M	SD	t
5 Component types' sequence	First element – plants	Control Experi.	1.00 2.78	.99 1.28	7.82***
	– others	Control Experi.	1.23 .36	1.16 .96	4.12***
	Intermediate elements – consumers	Control Experi.	3.35 4.98	2.83 2.21	3.25***
	– humans	Control Experi.	.08 .00	.33 .00	1.67*
	– decomposers	Control Experi.	.15 .00	.84 .00	1.31*
	Terminal element – decomposers	Control Experi.	2.27 3.16	2.73 2.90	1.60*
	– humans	Control Experi.	.19 .80	.86 1.96	2.03***
	– others	Control Experi.	.27 .26	1.17 1.19	.39
	Decomposers feeding on each element	Control Experi.	.29 1.42	1.19 2.35	3.05***
6 Components' hierarchical order	Random	Control Experi.	.00 .58	.00 1.64	2.50***
	Meaningful (size, strength, developmental stage)	Control Experi.	2.23 3.44	2.40 2.84	2.32***
7 Each element's number of functions	Single	Control Experi.	1.56 1.16	2.28 1.20	.90
	Single to multiple	Control Experi.	.44 .80	1.04 1.81	1.22
8 Matter characteritics	Type – organic	Control Experi.	.37 .94	.99 1.81	1.98**
	– inorganic	Control Experi.	.19 .48	.89 1.46	1.20*
	Conservation – conserved	Control Experi.	2.58 3.54	2.86 2.83	1.71
	– not conserved	Control Experi.	.00 .12	.00 .56	1.00*
	Transformation – transformed	Control Experi.	.38 2.68	.99 1.39	9.56***
	– untransformed	Control Experi.	.52 .76	1.31 1.10	1.00

Table continues on next page.

9	Energy characteritics	*Type*				
		– light	Control	.17	.92	4.74***
			Experi.	1.90	2.41	
		– heat	Control	.00	.00	2.19***
			Experi.	.28	.90	
		– life activities,	Control	.00	.00	3.06***
		metabolism	Experi.	.54	1.25	
		– chemical	Control	.56	1.64	6.87***
			Experi.	3.64	2.74	
		Conservation	Control	.46	.94	.66*
		– conserved	Experi.	.64	1.68	
		– not conserved	Control	.23	.61	4.51***
			Experi.	1.70	2.23	
10	Process characteritics	*Reversibility*	Control	.00	.00	1.86***
		– reversible	Experi.	.32	1.22	
		– irreversible	Control	2.38	2.79	1.12
			Experi.	1.78	2.68	
		Direction	Control	2.90	2.82	1.25
		– unidirectional	Experi.	3.62	2.95	
		– multidirectional	Control	.06	.24	.98*
			Experi.	.02	.14	
		Frequency/timing of	Control	2.58	2.91	4.14***
		events	Experi.	.60	1.82	
		– one at a time				
		– more than one	Control	.00	.00	2.57***
		simultaneously	Experi.	.06	.24	
11	Evolution	*Temporal*	Control	.13	.40	1.15*
		– short-term event	Experi.	.06	.24	
		– long-term event	Control	.31	.94	6.56***
			Experi.	3.04	2.80	
		Changes	Control	.19	.89	.19
		– individuals	Experi.	.16	.87	
		within lifetime				
		– gene frequencies	Control	.02	.14	2.74***
		in populations	Experi.	.78	1.96	
		Mutation control	Control	.23	.90	.41
		– by individuals	Experi.	.16	.87	
		– no control,	Control	.00	.00	2.89***
		random	Experi.	.80	1.96	
12	Chain length	*Unlimited* – cyclic	Control	.19	.52	1.95***
			Experi.	.04	.20	
		Limited	Control	.54	.50	3.75***
		– by element size,	Experi.	.20	.40	
		developmental				
		stage, strength				
		- by energy	Control	.06	.31	4.47***
			Experi.	1.06	1.56	

Table continues on next page.

Table 4.1. Continued

$* p < .05; ** p < .01; *** p < .001.$

Legend:

Biotic – a chain containing only biotic components.
Biotic and Abiotic – a chain containing both biotic and abiotic (earth) components.
Unidirectional – a chain always described as following a single direction – A feeds on B, which feeds on C, and so forth.
Multidirectional – a chain that may follow various directions (e.g., B feeds on C and is eaten by A).
Single function – elements in the chain model have a single function or characteristic (e.g., eater; producer, second consumer), usually combined with the unidirectional model.
Multiple functions – elements in the chain model may have several simultaneous functions or characteristics (e.g., a plant that is also a consumer), usually entailing a web model.
Linear – a chain model with an opening element and a closing one.
Cyclic – elements in the chain model are arranged in a circle where the "last" element is linked to the "first" one.
Branched – elements in the chain model feed on several additional elements that do not constitute elements in that chain (therefore linking to them like branches), usually found in combination with the unidirectional and single function models.
Combined branches – a model in which several chains are linked together but are still unidirectional; an element may feed on several elements from another chain but is not eaten by them.
Web – a model where many chains are linked through various elements in them; each may have several functions or characteristics; and any direction of links may be described.
Dynamic equilibrium – continuous molecular movement around a certain stable point, movement which may be described partly as reversible.

Hence, linking becomes inherent to the process of knowledge construction, as reflected in the significantly deeper, more holistic, and systemic comprehension of ecology evidenced in students after learning in the LTIL condition, as compared to the control group students.

The described analysis of students' responses to the probes enabled the construction of taxonomy for students' explanatory frameworks (Eilam, 2007b). The gross holistic changes that occurred in students' construct of feeding relations, in addition to the gradual specific changes regarding its various composites were probably instigated by the explicit instructional approach, the specially designed learning environment, and students' long engagement with problem solving situations. As suggested by Demastes et al. (1996), a desired cascade of associated concept changes is likely to be instigated following each single conceptual modification.

However, even in the LTIL condition, learning difficulties, mostly related to evolutionary as well as to biochemistry issues, persisted among students. These issues are not directly taught within the context of ecology, but it may be the case that by learning in a LTIL, the foundations are laid for their future successful acquisition. The traditional classroom lessons provided students with some understanding of the large principles of ecology and with training in their application in various episodes, thereby increasing students' awareness to their great complexity. Yet, only the LTIL pedagogy supplied students with opportunities to integrate the acquired theoretical knowledge into their own knowledge structures by continuously struggling to make sense of the constant flow of new data and dynamic changes in their ecosystems and to grant meaning to these theories through their incessant application while engaged in problem solving processes in their ecosystem. Many students were shown to be able to consider these episodes as parts of a whole ecosystem and as interrelated with other episodes and to various processes. Yet, this ability was constrained to a certain extent by some knowledge deficiencies in issues concerning energy, biochemistry, evolution, and the idea of simultaneity, as reflected in students' mental models of feeding relations.

Among other recommendations for learning environments designed to promote conceptual change, Vosniadou, Ioannides, Dimitrakopoulou, and Papademetriou (2001) suggested that curriculum coverage be narrowed to allow for a deeper exploration of fewer issues, as was manifested in the present leaning environment by the allocation of lengthy time for learning ecology. According to these authors, the sequence of acquiring the various systemic ecological concepts may be important only to the extent that the single concepts are fully acquired and understood before the holistic systemic view is presented and applied.

II. Small-Scale Hands-On Lab Experiment on the State of Matter

In the course of their inquiry, students performed many small-scale experiments designed to answer specific questions and thus to promote the inquiry. Often, the ideas for such experiments emerged from their own lines of thinking on the problems they were investigating in their ecosystems, but from time to time they originated from student-teacher discussions about problems students faced.

Studies have suggested that various learners may understand problems differently because they usually capture different features of the problem. This may result from (a) the nature of human perception, which captures some of a problem's features, misses others, and perceives some differently than they occur in reality; (b) learners' focus on features that are more salient or expected, granting them different weight (Anderson, Greeno, Reder, & Simon, 2000); (c) a problem's ambiguity, which may lead to differences in the perception of its features (Stanovich, 1999); (d) the human tendency to see patterns in situations where none exist and to make assumptions that are not stated or observed in the problem and then to go on to reason from them (Stanovich, 1999); and/or (e) learners' tendency to mix their held theories with their observations of evidence or data, thus to see and report what they expected rather than what they observed (Brewer, Chinn, & Samarapungaven, 2000; Gunstone & White, 1981). In addition, differences in perceiving a problem emerge from humans' tendency to contextualize problems with as much prior knowledge concerning the particular context as they can retrieve (Stanovich, 1999). The accessed prior knowledge stimulates the solver to attend to specific details of the observed phenomenon, grasp their importance, and grant them meaning accordingly (Simon, 2000). Manktelow (1999) suggested that prior beliefs might bring about a belief bias that results in the learner's focus of attention onto irrelevant task features and the learner's easier acceptance of information fitting those prior beliefs. All these factors probably impact learners' constructed knowledge representations from any experiments.

Although one of the studies in this research series focused on seventh graders learning ecology in a traditional mode (Eilam, 2004a), it demonstrated the limited effect of a small-scale experiment to instigate a process of conceptual change. This study described students' mental models of the structure of matter, as revealed in their responses while reasoning about an observed phenomenon in a small experiment they performed, namely, that in the exact same volume, a much larger number of soap water drops fall from a dropper than the number of water drops. Students' mental models revealed a static view of matter molecules. The application of this view resulted in many students perceiving the soap water to be denser. The students' reasoning (e.g., drops fell faster although no time measures were taken, gravitation was stronger for soap drops because more of them fell from the dropper) reflected many of the research findings described for students' understanding of the structure of matter. Most studies concerning conceptual change have suggested that a single lesson devoted to explaining an experiment and correcting various ideas may be insufficient to cause knowledge revision.

In contrast, when this same experiment was performed by one of the small groups in the context of the LTIL, a long process of conceptual change was activated. The teacher suggested this experiment as a small-scale examination of soap's effects on water characteristics (chemical, physical) in a group of students who were investigating the effects of detergents on their ecosystem's components and on its function as a whole. These students' initial responses were similar to those of the traditional classroom students. However, the LTIL students had the opportunity to experience a continuous examination of the phenomenon for as long as they wished, trying different liquids and examining the detergents' effects on plant growth and other measures before designing the full experiment that they followed for almost a year in their ecosystem. Gradually, the students constructed an understanding based on deep knowledge revisions over a long period of time. Such knowledge revision is quite rare in cases of a single detached experiment.

III. Skills Acquisition and Application

LTIL afforded a process of skill acquisition. As described above (see pp. 9-11, in the section describing the pedagogy applied students used a special textbook (Eilam & Aharon, 1998) for learning the inquiry skills.

Low-Order Skills and the Process of Constructing Procedural Knowledge

The video recordings of two student groups during one of the studies enabled the examination of students' acquisition of a simple skill that necessitated repeated applications over the course of the inquiry—the skill of choosing wisely (Eilam, 2002a). Students applied this skill whenever a single option had to be chosen from several alternatives (e.g., choosing a subject, a research question, a hypothesis, modes of experimentation, materials to carry it out), thus allowing for students' training in the application of this skill in many diverse new situations. However, unlike a regular case where a skill is learned and its application is trained through theoretical exercises, in the LTIL students applied it spontaneously as a tool they possessed to successfully advance their inquiry.

Students' discourse in each instance of application along the year was analyzed to reveal the process of constructing procedural knowledge. Ten explicit behavioral steps, which were generalized into three phases/levels, were identified in students' performance of the newly acquired skill, until they reached proficiency. This process of skill acquisition generally fit the processes described for declarative knowledge acquisition (Anderson, 1993; Korthagen & Lagerwerf, 1995; Shuell, 1990). Using Korthagen and Lagerwerf's ideas, the observed students might even be considered as having constructed a theory of choosing wisely in light of their ability to logically explain the skill's procedure, the importance and necessity of each of its operations, and the advantages in using it. These phases/levels of development were found in the case of choosing wisely to occur simultaneously with the former phase, being added while the former was still going on, rather than to occur sequentially. In the first phase, students encountered the new skill; clarified every word in the textbook description of its components as well as in the content in which it would be applied; and performed the relevant exercises. The students continued the clarification process as long as needed, as new applications were performed. The second phase began sometime after the initiation of the initial clarification phase and while it was still ongoing. Students constructed their understanding of core concepts of the skill, elaborated on them, and generalized them through new applications to be used in new situations. New dimensions were added to these concepts every time a new context was introduced. Students gradually learned to elicit only those criteria that differentiated among the options and were relevant to them. This phase continued in parallel with the first one as long as it was needed for internalizing new core concepts. In the third phase, students primarily performed the schema integration, automation, and manipulation of the skill, which evolved from an initial recognition of the interrelations among the isolated facts, information bits, and components (e.g., realizing the relation between choosing and ranking as expressed in the subjectivity of choosing). When such knowledge became automated to a certain degree, more mental resources were freed for the application of the skill in entirely new contexts. When students performed an automated application of the procedure, it occurred rapidly, with no textbook or teacher assistance, and the discourse revealed that students were able to think of other related issues rather than only of the operations, like of the meaning of the criteria in the new context.

Students' learning processes clearly demonstrated that their advancement through the various steps of the skill acquisition was unrelated to the specific content in which the application of the skill occurred. Rather, the number of times they applied the skill comprised the crucial factor, and that factor was enabled by the LTIL. Students progress was expressed in shifting (a) from a context-related behavior toward a decontextual behavior, (b) from specifics and book-related concepts and examples toward generalized and abstracted ones, (c) from mental resources invested in attention to the skill structure and textbook demonstrated applications toward lower requirements for mental resources by the automated knowledge application, and (d) from having declarative knowledge toward construction of procedural and conditional knowledge. These many shifts occurred along the time axis made possible by the LTIL. The LTIL exposed students to cross-situational experiences with the skill application, spread over the year. Each time they returned to the same skill, students possessed a more elaborated skill representation. The meaningful construction of this representation evolved from students' active participation, the relevance of the skill to the specific stage of the inquiry at the time, and its necessity for performing that stage (Eilam, 2002a).

High-Order Skills: Long-Term SRL and Metacognitive Thinking

The acquisition of high-order skills (e.g., inquiry, SRL) is basically similar to the acquisition of low-order skills but differs in several crucial respects. The composites of high-order skills cannot be presented as a list of operations to be carried out sequentially. High-order skill applications are non-linear in nature; therefore, such applications need to consider contextual factors, the content to be processed, prior knowledge and experiences concerning the skill and the content, as well as the management and orchestration of all involved operations and factors. A successful application sequence of the skill's various operations in a specific situation for achieving a specific goal cannot be predicted, resulting in feelings of uncertainty. The quality of the application products is determined, among other things, by the individual's ability to manage and orchestrate the application of these many operations (e.g., by an accurate definition of objectives, retrieval of the most appropriate skills for specific information processing in a specific situation, monitoring of progress) (Quellmalz, 1987). The LTIL enabled students to engage practically in the application of the complex high-order skills of SRL and metacognitive thinking (Eilam, 2007a; Eilam & Aharon, 2003; Eilam, Zeidner & Aharon, 2007). SRL is usually examined regarding short-term tasks. The LTIL revealed SRL-related behaviors that could probably not be revealed in short-term tasks. In short-term tasks, students may endure certain circumstances (e.g., a lag in progress) that cannot be ignored while performing a long-term task. Hence, the latter increases students' awareness of external and internal cues, which allow them to monitor their inquiries.

The LTIL structure challenged students (e.g., preserving motivation levels) and permitted them to overcome obstacles that, in a short-term task, might be considered a failure (e.g., being unsuccessful in one task but still succeeding in others). When tired, students in the LTIL could stop investing efforts for a while and catch up later with no negative repercussions (e.g., by assigning more homework, meeting in the afternoons), whereas during short-term tasks, such behaviors usually imply giving up. All of the students in the LTIL stayed on-task, at various levels of performance. Thus, the long-term project and its characteristics revealed many behaviors that might otherwise remain implicit to observers, suggesting that individuals may perform differently for short- and long-term tasks.

The analyses of the video recordings and of the yearly and daily reports (Eilam & Aharon, 2003) revealed eight SRL-related macrobehaviors among students while they independently planned, monitored, and readjusted their plans over time: (1) Getting into the habit of planning, which required consistency in completing the reports.

(2) Considering alternatives for achieving the set goals and making decisions concerning plans as afforded by the long sessions. Data showed that group members' emotions and social status might have interfered with this process, which involved argumentation and reasoning behaviors and the development of strategies for determining a preferred option. (3) Monitoring and reflection—awareness of one's own quality and quantity of knowledge, strategies, time, resources, and so forth. which is necessary for regulation, and was evident in students' reports. The gaps between the proposed and enacted plans increased students' awareness to the effectiveness of their activities and their selected strategies for achieving their stated goals, thus enabling them to change plans, become aware of their own learning preferences concerning a specific task, and even change their objectives if all efforts failed. (4) Increasing awareness of diverse external and internal cues over the course of monitoring, and the ability to act accordingly (e.g., task-related, strategy-related, environment-related, and personal cues). (5) Readjusting plans to improve rate of progress (e.g., mostly by assigning homework, extending available group time via extracurricular meetings, changing working habits). Such behaviors in a short-term task usually lead to quitting. (6) Demonstrating accountability for group decisions (e.g., sharing the load among group members, admonishing peers who did not follow group decisions). (7) Looking further ahead: planning for the future, and considering possible outcomes and conditions that might affect the inquiry. (8) Manipulating plans: intentionally setting a higher goal in order to increase motivation (e.g., by allocating a smaller amount of time than required for an activity, thus working faster and more efficiently).

Similarly to other reports (Alexander & Judy, 1988), students' accumulated experience with academic activity over time in the LTIL increased their knowledge about the task and influenced their SRL. Indeed, in the first few lab sessions, students invested most of their time and effort into managing the skill of filling out the reports rather than in the regulation process. In time, more of their mental efforts were allocated toward the regulation process as observed in their plans, choice of strategies, arguments about the optimal means to achieve goals and execute plans, and so forth. Students' quality and quantity of prior knowledge in the content domains and the required skill were found to be crucial. It permitted them to understand the task and its goals, to recognize the knowledge required for performing it, to perceive various relevant cues, and to predict performance. Consequently, prior knowledge facilitated students' capacity to monitor, plan accordingly, and judge outcomes with relation to goals, as well as to construct more appropriate conditional knowledge for better future performance (Butler, 1994; Butler & Winne, 1995).

The Eilam and Aharon (2003) study revealed four phases of students' SRL performance in the authentic ecosystem context. Phase 1 included forethought, planning, and activation—beginning each session with goal setting, planning, and enacting the plan. Phase 2 included perceiving cues from different sources and thereby performing monitoring. Phase 3 included control—acting on these cues by readjusting plans and manipulating them. Phase 4 included reaction and reflection—reflecting on activities and judging their efficiency in achieving the goals. These results match the SRL model suggested by Pintrich (2000), and suggest that basically, SRL applied for a long-term task is similar in nature to that applied while performing short-term tasks.

An additional study closely and qualitatively examined students' metacognitive thinking (Eilam, 2007a). Consistent use of the yearly and daily reports was found to promote students' construction and employment of metacognitive knowledge.

Finally, students' personality traits as predictors of their SRL behaviors and their achievements were examined (Eilam, Zeidner, & Aharon, 2007). Measurements of students' self-reported SRL were performed (using existing questionnaires) and were distinguished from measurements of students' enacted SRL, as assessed by the yearly and daily reports. Students' academic achievements were collected from school files. Data analysis (using regressions and path analysis) showed that students' enacted SRL, as measured by these tools, was the only direct predictor and that conscientiousness was the only personality trait that predicted academic achievements, mediated by the enacted SRL. The relation revealed between SRL performance and learners' personality suggests the need to design learning environments that account for the impact of students' dispositions on performance.

WHAT CAN BE DONE IN MORE TRADITIONAL LEARNING ENVIRONMENT?

As described in the chapter, the LTIL environment was long-term, rich with resources, and was designed to achieve several objectives in addition to improving students' understanding of ecology (e.g., acquisition of skills). However, the central idea was to enhance students' deep understanding of the systemic nature of ecosystems, by enabling them to manipulate variables in real live self-designed ecosystems and analyze the collected data. Such active learning promotes students' understanding of ways in which variables interact in systems and ways in which various factors may affect it. I believe it is possible to focus on achieving of the central idea in a more simplified LTIL.

First, if all students work with only a water ecosystem (rather than a greenhouse and a terrarium), teachers may be able to act more efficiently than when they have to disperse their efforts and resources over several kinds of ecosystems. A water ecosystem is also more easily recognized by students and simpler to manipulate. Students may choose organisms (e.g., fish, turtle, plants) for their own systems, thus still retaining a certain level of choice. Mostly, the "rich" resources used were (a) teachers' collections of articles, papers, videos recorded by teachers from the many TV programs dealing with ecology, pictures, and so forth, describing and discussing diverse issues in ecology; (b) simple chemical kits to measure variables in water or air, which are available science supply companies and may be prepared at school; (c) computers are not a must, if other information resources are available to students; (d) physics teachers may help by visiting the lab on specific occasions when energy (especially) is discussed, or integrate their instruction of the subject in their own lessons with the biology teacher; (e) although the long term lab provides students with opportunity to experience and practice independent self regulated learning (seldom available in schools), it is possible to promote students' awareness of the elements involved in self-regulation for short-term projects as well (e.g., stating goals, selecting strategies, monitoring for their effectiveness); (f) lessons may last 1½ or 2 hours rather than 3—as is usually the case in a school lab, and a part of it may be used for teaching ecological theories; and last (g) the teacher is the most important part of every learning environment and his/her way of conducting the LTIL is critical to its success. It is highly probable that students' learning of ecology can be enhanced in a less sophisticated LTIL.

CONCLUDING REMARKS:
IMPLICATIONS FOR THE TEACHING OF ECOLOGY

Complex inquiry environments allow for the development of many facets of learning, as well as a multiple-perspective exploration of phenomena. Students' increased familiarity with the environment and its phenomena through concrete interactions deepens their comprehension, increases its complexity, and permits them to formulate their knowledge in more abstract terms. This, in turn, allows students to move back and forth between the abstract and the concrete, between situated knowledge and knowledge removed from the situation (Roth, 1995).

The LTIL exposed students to ecological problems that challenged their intellectual abilities and could not be solved using preestablished formulas. Such experiences increase students' appreciation of the uncertainty involved in scientific activity (Gonzalez del Solar & Marone, 2001). Small-scale lab experiments that accompany traditional classroom learning lack this property. They usually demonstrate a limited phenomenon and thus may strengthen the linear instructional mode, therefore possibly hindering the understanding of ecology as a complex, systemic domain with non-linear interactions.

School experiences are multifaceted, and each classroom's occurrences achieve several goals concurrently. Some goals concerning the learning of ecology may be achieved by learning in the traditional mode (e.g., knowledge construction due to active involvement, scientific and cognitive skill acquisition, experiencing the scientific method). Participation in such short-term labs may bring about some other gains too (e.g., order, social construction of knowledge, collaboration). The LTIL was found to facilitate all of the above but may grant students with many more advantages that are considered by researchers and educators as of utmost importance. Among these, the opportunity to experience SRL, to acquire inquiry skills, or to develop deeper performance understanding of the systemic domain of ecology is prominent.

It is therefore highly recommended that students experience such an endeavor at least once at school, and if they do—let it be in ecology. The difficulties inherent to the learning of this domain, as well as the challenges of such a long-term responsibility, limit such learning mode to the junior high school and high school curricula. However, it is deeply believed by the author that some seeds of the required behaviors and experiences may already be planted successfully, with some modifications, even in preschool aged children.

ACKNOWLEDGMENT

The author would like to express her appreciation to Dee B. Ankonina for her editing contribution.

APPENDIX

THE APPLICATION OF THE SECOND LAW OF THERMODYNAMICS IN THE MEDITERRANEAN LOW FOREST

Objective: Possible Routes of Energy Transfer in a Natural Ecosystem of low Forest

Solar energy is transformed into chemical energy by the cells of leaves, stems, and green parts of the forest vegetation (grass, bushes, and oak, pine, and carob trees). Solar energy is invested in the chemical bonds of the organic matter, mainly sugars, produced from inorganic molecules that enter the plants cells via their roots (water) and their leaves' and stems' stomata openings (carbon dioxide). The organic molecules are produced through a multistage chemical process in these plant cells, in which chemical interactions among the various compounds occur. In this process, only part of the sun's energy is captured in the organic molecules. Some is lost as heat energy, which cannot be utilized by the cells' mechanisms. The rest is captured in the complex bonds of the organic molecules, which serve to sustain life. Cells' mechanisms were evolutionarily targeted to enable living processes. For example, certain types of molecules (mainly sugars and fats) evolved to release the energy tied up in their structure upon demand for fueling various cell processes. Fats, unlike sugars, yield the advantage of storage molecules that do not affect cell osmotic equilibrium. Other molecules evolved to serve as building materials (e.g., proteins, lipids, cellulose) for the cell organelles and structures, for its active molecules (the cell membranes, receptors, fibers, enzymes, vitamins, chlorophyll, hemoglobin), and for specific forest plant molecules like pine resin. The total energy utility of such processes is not high, due to the loss of heat at each point of transformation. At this point, one chemical compound changes into another, where the energy previously contained in it now becomes contained in a different molecule. Thus, all organic molecules—the pine resin, the flower nectar, or the proteins – contain energy. Some of the cell energy available to plants is invested in producing the next generation, and is contained in seeds, flowers, fruits, etc. that detach from the original plant. Some of the energy is contained in old parts of the plants, which are constantly being replaced, such as fallen leaves, epidermis, or various excretions like sap or nectar. Some of the energy is used for sustaining life (e.g., energy used for plant movement toward the light). Due to the energy loss and low utility of the mechanisms acting in biological cells, only that part of the energy that was invested in building certain compounds is available for the consumers feeding on the plants: herbivores and decomposers. Herbivores include birds who feed on forest seeds and nectar, rodents who feed on leaves, or insects who feed on plant materials. Decomposers may feed on dead plants' remains (like humus) or parts (fallen leaves, branches, seeds, etc.). Therefore, in order to supply their own demands for energy to sustain life, herbivores consume large amounts of the producers.

The story of energy transformation and how energy is utilized in cells repeats itself in the forest herbivores as well. Examples for the various modes in which energy is utilized in herbivores include: building molecules like protein and lipids; active and regulatory molecules like hormones, receptors, and enzymes; energy for movement—the contraction of muscle fibers; energy invested in the production of the next generation; or heat loss in the course of these many transformations.

Similar descriptions continue for the carnivores (birds of prey like falcons and eagles, predatory insects that feed on worms, snakes that feed on birds and small rodents) and for the decomposers. Decomposers may utilize all the energy contained in dead organic materials originating from the various feeding organisms along this route (plant bodies or body parts like fallen leaves or fruits; herbivores'/carnivores' carcasses, excretions, or dry epidermis). This idea of the loss of energy from one trophic level to another is presented in the model of the food pyramid.

REFERENCES

Alexander, P. A., & Judy, J. E. (1988). The interaction of domain-specific and strategic knowledge. *Review of Educational Research, 58,* 375-404.

Alparslan, C., Tekkaya, C., & Geban, O. (2003). Using the conceptual change instruction to improve learning. *Journal of Biological Education, 37*(3), 133-137.

Alters, B. J. (1999). What is creationism? *American Biology Teacher, 61,* 103-106.

Anderson, C., Sheldon, T., & Dubay, J. (1990). The effects of instruction on college nonmajors' conceptions of respiration and photosynthesis. *Journal of Research in Science Teaching, 27,* 761-776.

Anderson, J. R. (1993). *Rules of the mind.* Hillsdale, NJ: Erlbaum.

Anderson, J. R., Greeno, J. G., Reder, L. M., & Simon, H. A. (2000). Perspectives on learning, thinking, and activity. *Educational Researcher, 29*(4), 11-13.

Baron, J. B., & Sternberg, R. J. (Eds.). (1987). *Teaching thinking skills: Theory and practice* (pp. ix-xi). New York: W. H. Freeman.

Bentley, D., & Watts, M. (1992). *Communicating in school science: Groups, tasks and problem solving.* London: Falmer Press.

Bishop, B. A., & Anderson, C. W. (1990). Student conceptions of natural selection and its role in evolution. *Journal of Research in Science Teaching, 27*(5), 415-427.

Blackwell, W. H., Powell, J. M., & Dukes, G. H. (2003). *Journal of Biological Education, 37*(2), 58-67.

Brewer, W. F., Chinn, C. A., & Samarapungaven, A. (2000). Explanations in scientists and children. In F. C. Keil & R. A. Wilson (Eds.), *Explanation and cognition* (pp. 279-298). Cambridge, MA: MIT Press.

Brooks, J. G., & Brooks, M. G. (1993). *The case for constructivist classrooms.* Alexandria, VA: Association for Supervision and Curriculum Development.

Butler, D. L. (1994). From learning strategies to strategic learning: Promoting self-regulated learning by post secondary students with learning disabilities. *Canadian Journal of Special Education, 4,* 69-101.

Butler, D. L., & Winne, P. H. (1995). Feedback and self-regulated learning: A theoretical synthesis. *Review of Educational Research. 65*(3), 245-281.

Corno, L. (1993). The best laid plans: Modern conceptions of volition and educational research. *Educational Researcher, 22*(2), 14-22.

Cottrell, T. R. (2004). Difficult botanical concepts and previous knowledge. *American Biology Teacher, 66*(6), 441-445.

Crawford, B., Zembal-Saul, C., Munford, D., & Friedrichsen, P. (2005). Confronting prospective teachers' ideas of evolution and scientific inquiry using technology and inquiry-based tasks. *Journal of Research in Science Teaching, 42*(6), 613-637.

Demastes, S. S., Good, R. G., & Peebles, P. (1996). Patterns of conceptual change in evolution. *Journal of Research in Science Teaching, 33*(4), 407-431.

Driver, R., Asoko, H., Leach, J., Mortimer, E., & Scott, P. (1994). Constructing scientific knowledge in the classroom. *Educational Researcher, 23*(7), 5-12.

Driver, R., Leach, J., Millar, R., & Scott, P. (1996). *Young people's images of science.* Buckingham, England: Open University Press.

Driver, R., & Millar, R. (1986). Energy matters. Proceedings of an invited conference: *Teaching about energy within the secondary science curriculum.* Leeds, England: University of Leeds Center for Studies in Science and Mathematics Education.

Eilam, B. (2002a). Phases of learning: Ninth graders' skill acquisition. *Research in Science and Technology Education, 20*(1), 5-23.

Eilam, B. (2002b). Strata of comprehending ecology: Looking through the prism of feeding relations. *Science education, 86,* 645-671.

Eilam, B. (2004a). Drops of water and of soap solution: Students' constraining mental models of the nature of matter. *Journal of Research in Science Teaching, 41*(10), 970-993.

Eilam, B. (2004b). The instruction and acquisition of cognitive skills in Israel: Difficulties and accomplishments. In K. Mutua & C. Szymanski Sunal (Eds.), *Research on education in Africa, the Caribbean, and the Middle East* (Vol. 1, pp. 183-202). Greenwich, CT: Information Age.

Eilam, B. (2007a). *Promoting metacognition in the process of self-regulated learning: The case of 3 ninth grade students.* Manuscript submitted for publication.

Eilam, B. (2007b). *System learning and components understanding: A process of mutual enhancement.* Manuscript submitted for publication.

Eilam, B., & Aharon, I. (1998). *Science and inquiry.* Nazareth, Israel: El-Nahada Press.

Eilam, B., & Aharon, I. (2003). Students' planning in the process of self-regulated learning. *Contemporary Educational Psychology, 28,* 304-334.

Eilam, B., Zeidner, M., & Aharon, I. (2007). *Student conscientiousness, self-regulated learning, and science achievement: A prospective field study.* Manuscript submitted for publication.

Eilam, B. (Ed.), and Lilienfeld, L., Zernik, R., Carmon, H., & Rimon, S. (1980). *Man and landscape.* Haifa, Israel: University of Haifa, School of Education Curriculum.

Esiobu, O. G., & Soyibo, K. (1995). Effects of concept and vee mappings under three learning modes on students' cognitive achievement in ecology and genetics. *Journal of Research in Science Teaching, 32,* 971-995.

Fail, J. (2003). A no-holds-barred ecology curriculum for elementary and junior high school students. *American Biology Teacher, 65*(5), 341-346.

Finn, H., Maxwell, M., & Calver, M. (2002). Why does experimentation matter in teaching ecology? *Journal of Biological Education, 36*(4), 158-162.

Gallegos, L., Jerezano, M., & Flores, F. (1994). Preconceptions and relations used by children in the construction of food chains. *Journal of Research in Science Teaching, 31,* 259-272.

Gonzalez del Solar, R., & Marone, L. (2001). The "freezing" of science: Consequences of the dogmatic teaching of ecology. *Bioscience, 51*(8), 683-687.

Griffiths, A., & Grant, B. (1985). High school students' understanding of food webs: Identification of a learning hierarchy and related misconceptions. *Journal of Research in Science Teaching, 22,* 421-436.

Gunstone, R. F., & White, R. T. (1981). Understanding of gravity. *Science Education, 65,* 291-299.

Hogan, K. (2000). Assessing students' system reasoning in ecology. *Journal of Biological Education, 35*(1), 22-28.

Hogan, K., & Fisherkeller, J. (1996). Representing students' thinking about nutrient cycling in ecosystems: Bidimensional coding of complex topic. *Journal of Research in Science Teaching, 33,* 941-970.

Hatano, G., Siegler, R. S., Richards, D. D., Inagaki, K., Stavy, R., & Wax, N. (1993). The development of biological knowledge: A multi-national study. *Cognitive Development, 8,* 47-62.

Klein, J. (1990). *Interdisciplinarity.* Detroit, MI: Wayne State University Press.

Korthagen, F., & Lagerwerf, B. (1995). Levels in learning. *Journal of Research in Science Teaching, 32*(10), 1011-1038.

Krajcik, J., Blumenfeld, P., Marx, R. W., Bass, K. M., Fredericks, J., & Soloway, E. (1998). Middle school students' initial attempts at inquiry in project-based science classrooms. *Journal of Learning Sciences, 7*(3&4), 313-350.

Leach, J., Driver, R., Scott, P., & Wood-Robinson, C. (1996). Children's ideas about ecology 3: Ideas found in children aged 5-16 about interdependency of organisms. *International Journal of Science Education, 18,* 129-141.

Lee, C. A. (2003). A learning cycle inquiry into plant nutrition. *American Biology Teacher, 65*(2), 136-141.

Manktelow, K. (1999). *Reasoning and thinking.* East Sussex, England: Psychology Press.

McComas, W. F. (1998). The principal elements of the nature of science: Dispelling the myths. In W. F. McComas (Ed.), *The nature of science in science education: Rationales and strategies* (pp. 53-70). Dordrecht, the Netherlands: Kluwer Academic.

Minstrell, J., & van Zee, E. (Eds.). (2000). *Inquiring into inquiry learning and teaching in science.* Washington, DC: American Association for the Advancement of Science.

Munson, B. H. (1994). Ecological misconceptions. *Journal of Environmental Education, 24*(4), 30-34.

Ozay, E., & Oztas, H. (2003). Secondary students' interpretations of photosynthesis and plant nutrition. *Journal of Biological Education, 37*(2), 68-70.

Passmore, C., & Stewart, J. (2002). A modeling approach to teaching of evolutionary biology in high schools. *Journal of Research in Science Teaching, 39*(3), 185-204.

Perkins, D. N. (1987). Thinking frames: An integrative perspective on teaching cognitive skills. In J. B. Baron & R. J. Sternberg (Eds.), *Teaching thinking skills: Theory and practice* (pp. 41-61). New York: W. H. Freeman.

Perkins, D. N. (1992). *Smart schools.* New York: Macmillan.

Perkins, D. N. (1998). What is understanding? In M. S. Wiske (Ed.), *Teaching for understanding* (pp. 39-57). San Francisco: Jossey-Bass.

Perkins, D. N., & Salomon, G. (1989). Are cognitive skills context-bound? *Educational Researcher, 18*(1), 16-25.

Pickett, S. T. A., Kolasa, J., & Jones, C. G. (1994). *Ecological understanding: The nature of theory and the theory of nature.* San Diego: Academic Press.

Pintrich, P. R. (2000). The role of goal orientation in self-regulated learning. In M. Boekaerts, P. R. Pintrich, & M. Zeidner (Eds.), *Handbook of self-regulation* (pp. 451-502). San Diego, CA: Academic Press.

Pintrich, P. R., Marx, R. W., & Boyle, R. A. (1993). Beyond cold conceptual change: The role of motivational beliefs and classroom contextual factors in the process of conceptual change. *Review of Educational Research, 63,* 167-199.

Quellmalz, E. S. (1987). Developing reasoning skills. In J. B. Baron, & R. J. Sternberg (Eds.), *Teaching thinking skills: Theory and practice* (pp. 86-105). New York: W. H. Freeman.

Reiner, M., & Eilam, B. (2001). Conceptual classroom environment: A system view of learning. *International Journal of Science Education, 23*(6), 551-568.

Roth, W. M. (1995). *Authentic school science.* Dordrect, the Netherlands: Kluwer Academic.

Salomon, G., & Perkins, D. (1989). Rocky road to transfer: Rethinking mechanisms of a neglected phenomenon. *Educational Psychologist, 24,* 113-142.

Shuell, T. J. (1990). Phases of meaningful learning. *Review of Educational Research, 60*(4), 531-547.

Simon, H. A. (2000). Discovering explanations. In F. C. Keil & R. A. Wilson (Eds.), *Explanation and cognition* (pp. 21-59). Cambridge, MA: MIT Press.

Stanovich, K. E. (1999). *Who is rational? Studies of individual differences in reasoning.* Mahwah, NJ: Erlbaum.

Tessier, J. T. (2004). Ecological problem-based learning: An environmental consulting task. *American Biology Teacher, 66*(7), 477-484.

Unger, E. D., Seligman, N. G., & Noy-Meir, I. (2004). Grass grows, the cow eats: A simple grazing systems model with emergent properties. *Journal of Biological Education, 38*(4), 178-182.

Uyeda, S., Madden, J., Brigham, L. A., Luft, J. A., & Washburne, J. (2002). Solving authentic science problems. *Science Teacher, 69*(Jan.), 24-29.

Vosniadou, S., Ioannides, C., Dimitrakopoulou, A., & Papademetriou, E. (2001). Designing learning environments to promote conceptual change in science. *Learning and Instruction, 11,* 381-419.

Winne, P. H. (1995). Inherent details in self-regulated learning. *Educational Psychologist, 30*(4), 173-187.

Winne, P. H., & Perry, N. E. (2000). Measuring self-regulated learning. In M. Boekaerts, P. R. Pintrich, & M. Zeidner, (Eds.), *Handbook of self-regulation* (pp. 531-566). San Diego, CA: Academic Press.

Vygotsky, L. S. (1962). *Thought and languages.* Cambridge, MA: MIT Press.

Zimmerman, B. J. (1989). Models of self-regulated learning and academic achievements. In B. J. Zimmerman & D. H. Schunk (Eds.), *Self-regulated learning and academic achievement* (pp. 1-25). New York: Springer-Verlag.

Zimmerman, B. J., & Risemberg, R. (1997). Self-regulatory dimensions of academic learning and motivation. In G. D. Phye (Ed.), *Handbook of academic learning: Construction of knowledge* (pp. 105-125). San Diego, CA: Academic Press.

CHAPTER 5

STUDENTS' PERCEPTIONS OF THE SCIENCE LABORATORY LEARNING ENVIRONMENT

Sule Ozkan, Jale Cakiroglu, and Ceren Tekkaya

The purpose of the study was to examine seventh grade students' perceptions of the science laboratory environment and to investigate the association between laboratory learning and students' attitudes toward science and their science achievement. The study also investigated the differences in students' perceptions of science laboratory environment by gender. Data were collected with the Science Laboratory Environment Inventory and Science Attitude Scale. A total of 335 seventh grade students participated in the study. Results indicate positive and significant correlations between perceptions of the laboratory learning environment and the students' attitudes toward science and their science achievement. Gender was not found to be a significant factor that differentiates students' perceptions of laboratory learning environment.

INTRODUCTION

Over the past 30 years, considerable progress has been made internationally in terms of conceptualization, measurement, and

The Impact of the Laboratory and Technology on Learning and
Teaching Science K-16, pp. 111–134
Copyright © 2008 by Information Age Publishing
111

investigation of perceptions of the learning environment in science classrooms (Aldridge, Fraser, & Huang, 1999; Fraser, 1994; Goh & Fraser, 1998). Past research has mainly focused on the investigation of the association between students' cognitive and affective learning outcomes and their perceptions of psychosocial characteristics of their classrooms. Briefly, these studies suggest that a strong link exists between student outcomes and their perceptions of the learning environment (Fraser, 1994). These studies also noted that there are gender, subject, grade-level, school type, school location, and ethnic-related differences in classroom learning environments (Fraser, 1998; Fraser, 2003; Waldrip & Fisher, 2000).

ASSESSING STUDENT PERCEPTIONS OF SCIENCE LABORATORY ENVIRONMENTS: A LITERATURE REVIEW

A historical look at the field of the learning environment shows that a striking feature is the availability of a variety of economical, valid, and widely-applicable questionnaires for assessing student perceptions of classroom environments (Fraser, 1998). In the late 1960s and early 1970s, several of these questionnaires originated at about the same time in the research program of Herbert Walberg and Rudolf Moos. Walberg developed the Learning Environment Inventory (LEI) as a part of the evaluation process for the Harvard Project Physics (Walberg & Anderson, 1968). Moss designed social climate scales for hospitals and correction facilities which led to the development of the Classroom Environment Scale (CES) (Moos & Trickett, 1974). Other important learning environment instruments include the individualized classroom environment questionnaire (ICEQ) for open or individualized settings (Fraser, 1990), the College and University Classroom Environment Inventory (CUCEI) for higher education classrooms (Fraser & Treagust, 1986), the Science Laboratory Environment Inventory (SLEI) for laboratory settings (Fraser, Giddings, & McRobbie, 1995), the Questionnaire on Teacher Interaction (QTI) for assigning the interpersonal relationships between teacher and students (Wubbels & Levy, 1993), and the Constructivist Learning Environment Survey (CLES) for assessing the degree to which constructivist teaching and learning approaches are established in the classroom (Taylor, Fraser, & White, 1994). Among them, SLEI received much attention by the researchers due to the relationship between conducting laboratory activities and student achievement and attitude.

The laboratory is a unique learning environment which plays a vital role in science education. It is suggested that engaging learners in science laboratory activities promotes not only their understanding of scientific concepts and problem solving abilities but also their attitudes towards science (Hofstein & Lunetta, 1982; Lazarowitz & Tamir, 1994). Therefore, science educators need to obtain information and insight about what goes on in laboratory classes and what views students hold concerning their learning environment. For this purpose, Fraser, Giddings, and McRobbie (1995) developed the SLEI to assess students' perceptions of five subscales of classroom environment, that is, Student Cohesiveness ($\alpha = .78$), Open-endedness ($\alpha = .71$), Integration ($\alpha = .86$), Rule Clarity ($\alpha = .74$), and Material Environment ($\alpha = .76$) (see Table 5.1). *Student cohesiveness* measures the extent to which learners know, help, and are supportive of one another; *Open-endedness* measures the extent to which laboratory activities stress an open-ended, divergent approach to experimentation; *Integration* measures the extent to which laboratory activities are integrated with nonlaboratory and theory classes; *Rule clarity* assesses the extent to which behavior in the laboratory is guided by formal rules; and *Material Environment* assesses the extent to which laboratory equipment and materials are adequate (Fraser, Giddings & McRobbie, 1995).

The SLEI has been field tested extensively and cross validated in different countries with a sample of 3727 senior high school and

Table 5.1. Descriptive Information for Five Subscales of the SLEI (Adapted from Fraser, Giddings, & McRobbie, 1995)

Subscale	Subscale Description	Sample Item
Student Cohesiveness (SC)	The extent to which students know, help, and are supportive of one another.	I get on well with students in this laboratory class.
Open-Endedness (OE)	The extent to which the laboratory activities emphasize an open-ended, divergent approach to experimentation.	In this laboratory class, I am required to design my own experiments to solve a given problem.
Integration (I)	The extent to which the laboratory activities are integrated with nonlaboratory and theory classes.	I use the theory from my regular science class sessions during laboratory activities.
Rule Clarity (RC)	The extent to which behavior in the laboratory is guided by formal rules.	My laboratory class has clear rules to guide my activities.
Material Environment (ME)	The extent to which the equipment and materials are adequate	The equipment and materials that I need for laboratory activities are readily available.

university students in 198 science laboratory classes in six countries (Australia, United States, Canada, England, Israel, and Nigeria). Since then, several researchers have used the SLEI to assess student perceptions of laboratory learning environments and attempted to identify the relationships between those perceptions and student attitudinal and cognitive outcomes (Giddings & Waldrip, 1996; Henderson, Fisher, & Fraser, 2000; Lee & Fraser, 2001; Wong & Fraser, 1994). For example, Wong and Fraser (1994) investigated associations between 10th-grade Singaporean students' (n = 1592) perceptions of their chemistry laboratory classroom environment and their attitudes towards chemistry. Results of the study indicated that all five SLEI subscales were associated significantly with attitude. Multiple regression analysis indicate that the subscales integration and Rule Clarity were the strongest and most consistent predictors of the attitude. Wong and Fraser concluded that chemistry laboratory lessons which integrate knowledge learned from regular classroom chemistry lessons and provide clear rules for students to follow have a positive effect on the students' chemistry-related attitude. In addition, Wong and Fraser also noted that there was a negative association between Open-endedness and attitude to scientific inquiry in chemistry. This suggests that Singaporean students' attitudes towards accepting scientific inquiry as a way of thought in chemistry were more positive in chemistry laboratory classes which favored close-ended activities. With respect to gender, their findings replicated those of some previous studies in that female students were found to perceive their laboratory environment more favorably than did male students.

Using SLEI Giddings and Waldrip (1996) conducted a comparative international study to investigate secondary schools (year 10 and 11) students' (n = 7786) science laboratory learning environment across both developed and developing countries (Australia, Brunei, Cook Islands, Fiji, Papua New Guinea, Singapore, Solomon Islands, Tonga, Tuvalu, United States, Vanuatu, and Western Samoa). Results revealed that participants throughout the countries generally viewed their science laboratory environment favorably, but there was a low perception on the Open-endedness scale.

In another study, Hofstein and Cohen (1996) compared the Israeli high school students' perceptions of laboratory learning environment in chemistry and biology by using the actual and preferred form of the SLEI. A total of 371 11th-graders participated in their study. Using Hotelling's T^2 statistics, significant differences between chemistry and biology laboratory environments were found on two subscales of the SLEI, namely integration and Open-endedness. Similarly, differences between biology and chemistry students were found on both the actual and preferred forms of the integration, Open-endedness, and Rule Clarity subscales. Hofstein and Cohen concluded that the SLEI instrument is sensitive to different science curricula and different instructional methods utilized in science laboratories. Lee and Fraser (2001) examined the Korean high school students' perceptions of laboratory learning environments and their attitudes towards science by using the SLEI and the Test of Science-Related Attitudes (TOSRA). The study involved 437 students from three different streams: humanities stream, science-oriented stream, and science-independent stream. Lee and Fraser reported that Korean students viewed their science laboratory favorably. For example, the Korean students perceived a relatively high level of cohesiveness in their laboratory lessons. The students reported that their laboratory classes were highly coordinated with theory classes (regular science classes) and the rules in laboratory classes were clear. However, the materials and equipment were perceived to be inadequate. This study also reported low level of Openendedness in the laboratory environment. When the perceptions of students across the streams were compared, it was reported that students from the science-independent stream perceived their laboratory environments more favorably than did students in the other streams. This study also demonstrated associations between laboratory learning environments and their attitudes towards science. Simple correlations showed that there was a statistically significant correlations between each of the four TOSRA scales and the SLEI scales of students cohesiveness, integration, and Rule Clarity.

Research by Henderson, Fisher, and Fraser (2000) in secondary biology laboratory classrooms in Tasmania, Australia, also replicated earlier findings of positive relationships between student outcomes and the science laboratory environment. A sample of 489 students from 28 senior biology classes in eight schools completed the QTI, the SLEI, and two attitude questionnaires (Attitude to This Class and Attitude to Science Laboratory Work). Simple correlation analysis indicates statistically significant associations between attitudinal outcomes and all SLEI scales except Open-endedness. Multiple regression analysis indicates that integration was the scale most strongly associated with attitudinal outcomes when other SLEI scales were mutually controlled. Furthermore, many aspects of interpersonal teacher behavior and the laboratory learning environment were associated with students' attitudinal outcomes.

For example, favorable student attitudes were found to be associated with students' perceptions of the teacher's strong leadership behavior, a greater degree of integration of practical and theory work, and more Rule Clarity. Furthermore, it was found that the teacher's strong leadership (e.g., setting tasks, holding students' attention), provision of a degree of student responsibility and freedom (e.g., giving the freedom and responsibility to students for their own activities), and integration of practical and theory components of the course were likely to promote achievement. However, a greater degree of teacher control of learning, emphasis on Rule Clarity, and an open-ended approach to the course were negatively associated with student achievement. Henderson, Fisher, and Fraser (2000) also found that the students' perceptions about the integration of laboratory activities and theory classes were positively related to their achievement, whereas a greater degree of emphasis on Rule Clarity and an open-ended approach to laboratory was negatively associated with student achievement.

Recently, Tsai (2003) explored the differences between Taiwanese science students' and teachers' perceptions of laboratory environments using SLEI. About 1,012 eight and ninth grade students and 24 science teachers participated in the study. According to the results of the t test, students perceived their actual laboratory environments as less open-ended, less integrated with theory class, less student-cohesive, and with less Rule Clarity than they preferred. Compared to their teachers, students showed much more dissatisfaction with approaches to laboratory activities. For example, students prefer a laboratory learning environment where they have more student cooperation, conduct more open-ended investigations, and explore more deeply into the connections between theory and practical evidence, while having clearer rules for guidance and better material support than their teachers. The teachers in the study, however, showed stronger preferences for the support of better equipment and material for laboratory work than their students. This study suggested that epistemological views about science might be one of the important factors causing differences between students and teachers of laboratory learning environments. According to interview results, the Taiwanese teacher thought that the laboratory was best used to obtain accurate and almost certain results to verify the well-known scientific knowledge. Tsai claimed that these ideas indicated that epistemological views of teachers about science were more oriented to positivism and empiricism and accordingly guided their perceptions of the goals of laboratory activities.

Hofstein, Nahum, and Shore (2001) explored the idea that the science laboratory provides a unique learning environment that differs from the learning environment that exists in classrooms in which other instructional techniques are employed. In their study, SLEI was used to assess the students' perceptions of their chemistry laboratory learning environment. The sample consisted of two groups of students, the inquiry and the control groups. The inquiry group consisted of 130 11th-grade students and the control group consisted of 185 11th-grade students. T test analysis showed significant differences between the groups concerning their actual perceptions. The study results also demonstrated that the differences between the actual and preferred laboratory learning environment were significantly smaller for the inquiry group than for the control group.

Research on gender differences in classroom environment perceptions was also conducted in various countries (Fraser, Giddings, & McRobbie, 1995; Henderson, Fisher, & Fraser, 2000; Quek, Wong, & Fraser 2002). Working with 497 senior high school chemistry students from 18 classes in Singapore, Quek, Wong, and Fraser (2002) investigated gender differences in the students' perceptions of their chemistry learning environments using Chemistry Laboratory Environment Inventory, evolved from SLEI. Results of ANOVA indicated significant differences between the boys' and girls' perceptions of their chemistry laboratory classroom environment in favor of girls. As far as students responses to subscales are concerned, Quek, Wong, and Fraser reported that the greatest gender difference in actual perceptions were found for the Material Development scale, implying that boys perceived their actual chemistry laboratory classroom environment to be much less well equipped than did girls. In addition, boys perceived fewer rules and restrictions being imposed in their chemistry laboratory classroom environment. The authors concluded that boys were generally more task oriented than girls and, thus, paid less attention to the physical environment. Girls, however, appeared to perceive a low level of cohesiveness among their classmates compared to boys. The findings also showed that girls preferred to learn using more open-ended activities in a more organized and well equipped chemistry laboratory environment.

Some prior research in Turkey shows similar findings regarding the perception-attitude relationship in the Turkish context. For example, Telli, Cakiroglu, and den Brok (2006) conducted a study to investigate Turkish high school students' perceptions and their attitude toward biology. Perception data were gathered with 1,983 9th- and 10th-grade students from 57 biology classes at schools in two major Turkish cities. Data were collected with an adapted and translated version of the What is Happening in This Classroom (WIHIC) instrument and the Test of Science Related Attitudes (TOSRA). Correlation and regression analyses revealed that students' perceptions of their learning environment in biology were significantly associated with their attitudes.

In another study, Cakiroglu, Tekkaya, and Rakici (2007) investigated two aspects of classroom environment, namely learning environment and interpersonal teacher behavior in science classrooms. Data were collected from 722 eighth grade students in 24 conveniently selected classes in five schools in large urban district of Ankara, Turkey, using the WIHIC, the QTI, and the Science Attitude Scale. Results indicated that the students generally perceived a positive science classroom learning environment.

Students also perceived that their science teachers run their classes with fairly strong leadership, accompanied by a helping/friendly, understanding behavior with a fairly strict behavior. Turkish students felt that their teachers were less uncertain, dissatisfied, and admonishing. Also a positive relationship between the students' perception of their science teachers' interpersonal behaviors and learning environment and their affective outcomes was found. Girls perceived their learning environments positively and viewed their teachers as displaying more leadership, helping/friendly, and understanding behaviors on the Questionnaire on Teacher Interaction than do boys.

Recently, Arisoy, Cakiroglu, and Sungur (2007) examined the relationship between school students' perceptions of science classroom environment from constructivist perspective and their adaptive motivational beliefs. A sample of 956 eight grade students in 36 elementary science classes in Turkey completed Turkish versions of the CLES, and Motivated Strategies for Learning Questionnaire (MSLQ). Results of a canonical analysis showed that perception of higher levels of five key elements of critical constructivist learning environment—personal relevance, uncertainty, critical voice, shared control, and social negotiation—were associated with higher levels of students' adaptive motivational beliefs. This finding suggests that favorable student motivation could be increased in classes where students perceive more personal relevance, shared control with their teachers, freedom to express concern about their learning, science as ever changing, and interact with each other to improve comprehension.

Literature Review Summary

To sum up, review of research on assessing student perceptions of science laboratory environments (see Table 5.2) revealed that students from different countries and from different streams, generally perceived their science laboratory learning environment positively. Studies examining the relationships among students' perceptions of science laboratory environments, science achievement, and their attitudes towards science demonstrated positive associations between these variables. These studies suggested that students holding positive perceptions about their science laboratory learning environment also have favorable attitude toward science and high science achievement scores. As far as gender is concerned, it was reported that females and males were differed with respect to their perceptions of science laboratory environments. Generally, females perceived their laboratory environments more favorably compared to males.

Table 5.2. Assessing Students' Perceptions of Science Laboratory Environments

Authors and Date	Title of Article	Variables Compared	Instruments Used	Key Conclusions
Wong & Fraser (1994)	Science laboratory classrooms environments and students attitudes in chemistry classes in Singapore.	Perceptions of chemistry laboratory classroom learning environment	Chemistry Laboratory Learning Environment Inventory (CLEI) (a modified form of SLEI)	Teachers' and students' perceptions were different.
		Attitude towards chemistry		Positive relationship was found between laboratory environment and attitude.
		Gender	Questionnaire of Chemistry-Related Attitudes (QOCRA)	Female had more favorable perceptions than males.
Hofstein & Cohen (1996)	The laboratory environment of high school students in chemistry and biology laboratories.	Perceptions of actual and preferred chemistry and biology classroom learning environment	Science Laboratory Learning Environment Inventory (SLEI)	Significant difference was demonstrated between chemistry and biology laboratory learning environment in Integration and Open-Endedness subscales.
		Gender		Chemistry and biology students held different perceptions of actual and preferred forms of the SLEI with respect to Integration, Rule Clarity and Open-Endedness subscales.
				Girls had more favorable perceptions in the actual form of SLEI than the boys.

Table continues on next page.

Table 5.2. Continued

Authors and Date	Title of Article	Variables Compared	Instruments Used	Key Conclusions
Giddings & Waldrip (1996)	A comparison of science laboratory classrooms in Asia, Australia, South Pacific and USA: An international study.	Perceptions of Science laboratory classroom learning environment	Science Laboratory Learning Environment Inventory (SLEI)	Participants viewed their science laboratory environment favorably, but there was a low perception on the open-endedness scale.
		Attitude towards science	Attitudes Questionnaire	
		Gender		Male students' attitudes towards science were more favorable than females.
Lee & Fraser (2001)	Laboratory classroom environments in Korean high schools.	Perceptions of science laboratory learning environment	Science Laboratory Learning Environment Inventory (SLEI)	Relationship between laboratory classroom environment and students' attitudes towards science were found.
		Attitude towards science	Test of Science-Related Attitude (TOSRA)	

Study	Focus	Variables	Instruments	Findings
Henderson, Fisher, & Fraser (2000)	Interpersonal behavior, laboratory learning environments, and student outcomes in senior biology classes	Perceptions of science laboratory learning environment	Science Laboratory Learning Environment Inventory (SLEI); the Questionnaire on Teacher Interaction (QTI)	A strong association was found between students' perceptions of learning environment and attitudinal outcome.
		Attitude	Attitude to Science Laboratory Works (adapted from TOSRA)	Certain aspects of learning environment were found to be related with cognitive and practical performance outcomes.
		Achievement	Attitude to This Class; Practical Test Assessment Inventory (PTAI)	Unique contributions of the QTI and SLEI to variance in achievement were indicated.
Quek, Wong, & Fraser (2002)	Gender differences in the perceptions of chemistry laboratory classroom environments	Perceptions of actual and preferred chemistry laboratory classroom learning environment	Chemistry Laboratory Learning Environment Inventory (CLEI)	Significant differences between perceptions of boys and girls were found.
		Gender		Girls perceived their laboratory learning environment more favorable than boys.

AN INVESTIGATION OF PERCEPTIONS OF SCIENCE LABORATORY LEARNING ENVIRONMENT AND THEIR ASSOCIATION WITH SCIENCE ATTITUDES AND ACHIEVEMENT

The domain of learning environments research is a new research field in Turkey, and few related works are available. This study was the first in its kind to connect the SLEI to student attitudes in Turkey and thereby may add insight to this field in the region. The present study investigates associations between students' perceptions of their laboratory environment, students' attitudes, and science achievement. As used in this study, science achievement was defined as the school science performance of the students as indicated by their science course grades. Attitude toward science was considered as the feelings, beliefs, and values held about the enterprise of science as learned in school science classes, science in general, and the effect of science on a society (Ousbourne, Simon, & Collins, 2003). Students' perceptions of their laboratory environment, as used in this study, were defined in terms of Student Cohesiveness, Open-endedness, Integration, Rule Clarity, and Material Environment.

Specifically, this study focused on the following research questions:

1. What are the relationship between students' perceptions of their laboratory learning environment, their attitudes toward science, and their science achievement?

2. Are there significant differences in students' perceptions of their laboratory learning environment by gender?

Study Setting

In Turkey, science laboratories generally have some similar physical characteristics. The physical settings of the science laboratories, as indicated in Figure 5.1, are characterized by benches surrounded by chairs which allow students to work cooperatively in small groups throughout the investigation process. The Material Environment in the science laboratories shows variation from school to school. Laboratory equipment can be divided into major groups like glass materials (flasks, beakers, funnels, graduated cylinders, Petri dishes, etc.), chemicals, mechanical instruments (balance, scale, etc.), and materials required for physics, or other science experiments such as electricity and magnetism (see Figure 5.1). In addition, the science laboratory environment is enriched by a variety of charts and models about the human body, animals, plants, earth science, and so forth. (Figure 5.2 and Figure 5.3).

Figure 5.1.   Materials for laboratory equipment.

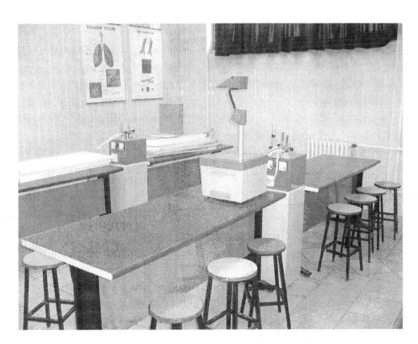

Figure 5.2.   Physical Setting of a science laboratory in Turkey.

Figure 5.3.   Laboratory equipment cases with examples of science models.

Generally, science laboratories also provide for use of overhead or other projectors. Regarding science laboratory instruction, teachers generally prefer a traditional, direct teaching approach while performing experiments. Students, in this environment, were expected to follow ready made experimental procedures and were directed to reach a result about scientific phenomena.

Study Design

Sample

The participants of this study were 335 seventh grade students from three schools in a large district of Ankara, the capital of Turkey. Of the 335 students in the sample, 168 were girls and 167 were boys. Schools in the present study were in the same large school district and were comparable in terms of student age, overall grade-point average in science, and socioeconomic status. The socioeconomic status of the students was similar, with the majority of the students coming from middle- to high-class families. The educational level of the parents can also be considered as high with 36% of the parents graduated from high school and 38% graduated from university. The school settings have similar characteristics as well. Class size in these schools varied from 30 to 40 students. The schools have a science laboratory which has enough space and materials for doing various science experiments.

Instrumentation

The data collected from students included two kinds: (1) responses to the SLEI, and (2) responses to the Science Attitude Scale. The questionnaires were administered to the participants after getting permission from the administration. Then, in each class, students were told about the purpose of the questionnaire and the procedure for completing it. After this short explanation, students were asked to complete the questionnaire on their own. They were instructed to think about each question and answer it as it applies to them. It took about one class hour for the students to complete the questionnaire.

Science Laboratory Environment Inventory (SLEI)

The SLEI, developed and validated by Fraser, Giddings, and McRobbie (1995), was administered to investigate students' perceptions of science laboratory environments. This questionnaire has 35 items in 5 different subscales (Student Cohesiveness, Open-Endedness, Integration, Rule Clarity, and Material Environment) including 7 items in each subscale. It consists of items with a 5-point Likert-type response scale with the following alternatives: (5) Almost Always, (4) Often, (3) Sometimes, (2) Seldom, and (1) Almost Never. In its original form, this inventory has two questionnaires, one of which investigates the students' views about actual laboratory environments (called "Personal Actual Form"), and the other assesses students' perceptions of ideal laboratory environments (called "Personal Preferred Form"). In this study, only the Personal Actual Form was used. Descriptive information about the subscales of the SLEI and representative/sample items for the subscales was presented in Table 5.1.

Science Attitude Scale

Students' attitudes toward science were assessed using the Science Attitude Scale developed by Geban, Ertepinar, Yilmaz, Altin, and Sahbaz (1994). This scale consists of 15 items and is designed to be rated on a 5-point Likert type response format (strongly disagree, disagree, neutral, agree, and strongly agree). The alpha reliability coefficient of the questionnaire was found to be 0.93 suggesting that the instrument has adequate internal consistency reliability.

Data Analysis

There were three distinct components of data analysis conducted in the current study. First, data from the SLEI were subjected to Confirmatory Factor Analysis (CFA) and the subscales were validated in terms of factor structure and reliability. The second component of data analysis involved using simple correlations to investigate the associations between perceptions of laboratory learning environment and students' outcomes (achievement and attitudes). Third, MANOVA was used to identify differences in students' perceptions of laboratory learning environments across gender. In this analysis, gender was considered as independent variables and each dimension of SLEI scores were considered as the dependent variables. The analysis was performed with the significance level of $\alpha = 0.05$ using SPSS. Data about students' science achievement was obtained by collecting information about science grades of students.

Results

Validation

First, the SLEI was translated and adapted into Turkish by the authors. The next step involved an independent back translation of the Turkish version into English by a qualified, bilingual Turkish instructor. The back-translation was reviewed and modifications in wording for the Turkish translation were carried out. The modified version of the SLEI was piloted with a group of students. The item validity and the comprehensiveness of the statements were evaluated and all necessary changes were made. The reliability coefficients of the subscales as estimated by Cronbach's alpha range from .50 to .72 with a total scale reliability of .85, suggesting that the final instrument had adequate internal consistency.

Confirmatory Factor Analysis

A one-factor model was proposed to be confirmed in this part of the analysis. This model was fitted within the sample by using the confirmatory factor analysis (CFA). The fit indexes to be used for evaluating the model proposed were goodness of fit index (GFI), adjusted goodness of fit index (AGFI), root mean square error of approximation (RMSEA), and standardized root mean square residuals (S-RMR). Values of 0.08 or less in RMSEA and SRMR claim a good model data fit (Schreiber, Stage, King, Nora, & Barlow, 2006). As seen in Table 5.3, these indices indicated a reasonable fit of the proposed model. However, GFI and AGFI should be above 0.90 to be regarded as a close fit (Schreiber et al., 2006). Considering the values obtained for the RMSEA and SRMR as adequate, the researchers decided to continue the analysis with this model.

Table 5.3.   Goodness-of-Fit Indices

Indices	Values
GFI	0.82
AGFI	0.86
RMSEA	0.074
S-RMR	0.08

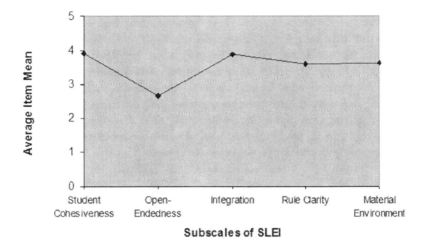

Figure 5.4.   Students' average perceptions on the SLEI.

The analysis demonstrated that the translated and adapted version of the SLEI was a reliable and valid instrument for assessing Turkish students' perceptions of their science laboratory environment.

Descriptive Statistics Results

The science laboratory environment was described by students' responses to the SLEI. Figure 5.4 (above) provides a graphical representation of the mean scores for each subscale. The results indicate that students perceived their science laboratory environment relatively favorably, as indicated by the mean scores ranging form 2.66 to 3.91 on a 5-point scale. For example, the mean score for the Student Cohesiveness ($M = 3.91$; $SD = .66$) subscale indicate that students perceived a high level of cohesiveness in their laboratory sessions. The mean for this subscale was the highest of all subscales, indicating that students know, help, and are support one another in the laboratory. The integration subscale had the next highest mean ($M = 3.89$; $SD = .71$), implying that laboratory activities were integrated with nonlaboratory and theory classes. The mean values for the Rule Clarity ($M = 3.61$; $SD = .64$) and Material Environment ($M = 3.63$; $SD = .74$) subscales were moderate. Students perceived that the rules in laboratory classes are relatively clear, suggesting that behavior in the laboratory is likely to be guided by formal rules. Similarly, material and equipment in the laboratory were perceived to be adequate. The lowest mean score appeared for the Open-Endedness subscale, ($M = 2.66$; $SD = .61$) implying that the laboratory activities do not necessarily emphasize an open-ended, and divergent approach to experimentation.

Descriptive statistics for science achievement and science attitude scores of the three schools involved in the study are presented in Table 5.4. Results indicate that science achievement mean scores of the three schools were quite similar to each other, two schools (school one and school two) have exactly the same mean values. Considering the possible maximum achievement score, it can be concluded that students' science achievement in the selected schools were slightly above average. Regarding the science attitude mean scores, school two had the highest mean value ($M = 57.96$, $SD = 13.29$), followed by school three ($M = 53.21$, $SD = 14.85$), and then by school one ($M = 49.91$, $SD = 13.65$). It can be concluded that students' attitudes toward science were above the avarege approaching the maximum end of the possible range.

Simple Correlations Results

The results of the simple correlation analysis, see Table 5.5, demonstrated that four among five subscales of the SLEI were significantly and positively correlated to students' science achievement. As indicated in

Table 5.4. Descriptive Statistics for Science Achievement and Science Attitude Scores of the Three Schools

		Mean	SD	Minimum	Maximum
Science achievement	School 1	3.05	1.47	1	5
	School 2	3.02	1.46	1	5
	School 3	3.05	1.31	1	5
		Mean	SD	Actual Range	Possible Range
Science attitude	School 1	49.91	13.65	16-75	15-75
	School 2	57.96	13.29	18-75	15-75
	School 3	53.21	14.85	17-75	15-75

Table 5.5.   Simple Correlations With Two Outcome Measures

		SC	OE	I	RC	ME
Science achievement	Pearson correlation	.284**	−.033	.287**	.167**	.142*
	% Variance accounted for	8	0	8	3	2
	Effect size	.28	.03	.28	.17	.14
Attitude	Pearson correlation	.313**	.217**	.447**	.400**	.398**
	% Variance accounted for	10	5	20	16	16
	Effect size	.31	.22	.45	.40	.40

** Correlation is significant at the 0.01 level (2-tailed).
* Correlation is significant at the 0.05 level (2-tailed).

Table 5.5, Pearson product moment correlation coefficients ranged from 0.14 to 0.28. The highest correlation to science achievement were the scores for the Integration subscale, which demonstrated that to the greater extent that students perceived their laboratory activities are integrated with theory classes, the higher their science achievement scores were found. Similar associations were found between science achievement and the perceptions of student cohesiveness, rule clarity, and material environment subscales. Effect sizes for each correlation coefficient and percent of variance of science achievement accounted for by its linear relationship with each predictor variable were also presented in Table 5.5. Correlation coefficients were used as an indication of effect size. According to Green, Salkind, and Akey (2000), correlation coefficients of .10, .30, and .50, regardless of sign, are interpreted as small, medium, and large effect size respectively.

As demonstrated in Table 5.5, all five subscales of the SLEI were found to be significantly and positively correlated with students' attitudes towards science. Those correlations range from 0.21 to 0.45. As in science achievement, the highest correlate to attitude towards science was the scores for the Integration subscale. Students' perceptions about the integration of laboratory activities with their theory classes correlated, significantly, with their attitude towards science. Overall, these results indicate positive and significant associations between perceptions of the laboratory environment and the students' attitudes toward science and their science achievement. Effect sizes for each correlation coefficient and percent of variance of science attitude accounted for by its linear relationship with each predictor variable were also presented in Table 5.5.

Multivariate Analysis of Variance Results

A one-way MANOVA was conducted to determine gender differences on the five dependent variables, Student Cohesiveness, Open-Endedness, Integration, Rule Clarity, and Material Environment. MANOVA results comparing males and females revealed that there was no overall significant difference by gender on the collective dependent variables, Wilks' Lambda = .98, $F(5,329) = 1.57$, $p = .17$, $\eta^2 = .023$.

Discussion

This study provided insights into the science laboratory learning environment in Turkey through the eyes of the students. The findings indicated that Turkish seventh grade students in the selected settings show relatively favorable perceptions of their laboratory learning environment, with the lowest score occurring for the Open-Endedness subscale. It appears that participants perceive their science laboratory learning environment in such a way that it integrates knowledge learned from theory classes, offers opportunity to support each other, and provides clear rules and materials. In addition, the students reported that actual laboratory learning environments did not greatly emphasize an open-ended approach to experimentation. This finding was consistent with the results reported by previous studies (Giddings & Waldrip, 1996; Lee & Fraser, 2001; Wong & Fraser, 1994).

A possible explanation for the low correlation of open-ended laboratory work might lie in the fact that there is a highly centralized and prescriptive mode of education in Turkey (Cakiroglu, Telli, & Cakiroglu, 2003). The laboratory approach in these schools was oriented in a traditional manner. In other words, science laboratory lessons were generally performed using prescribed procedures and did not emphasize an open-ended, divergent approach to experimentation. This finding might be also attributed to the presence of examination-oriented school system in Turkey. The students who graduate from 8 years of elementary education may take the Secondary Education Selection and Placement Examination to enroll in selective high schools or can continue their studies at a general secondary school, depending on the graduate's grades of elementary education. Both students and teachers do not want much time spent on an open-ended approach to experimentation in laboratory because they believe laboratory lessons emphasizing open-ended approach do not guarantee student achievement in the exam (Cakiroglu, Telli, & Cakiroglu, 2003). Therefore, these results support past research studies which found science laboratory classes are dominated by closed-ended activities (e.g., Giddings & Waldrip, 1996; Lee & Fraser, 2001; Wong & Fraser, 1994).

Concerning the SLEI Integration subscale, finding higher mean scores is not surprising because laboratory activities in Turkey are performed mainly to clarify the theories presented in the science classes and in the science textbooks. Traditional laboratory activities provide opportunity to present the theories in a more concrete way. Traditional approaches to laboratory experiments are conducted to verify the knowledge provided by teacher in the classroom. In a similar fashion, regarding the high rating for Cohesiveness subscale, participants were given an opportunity to work in groups in science laboratories which encourages them to work cooperatively with and support one another during laboratory activities.

Regarding the associations between students' perceptions of their science laboratory environment and their science attitudes, all five subscales of the SLEI were found to be positively related to students' attitudinal outcomes. Students' attitude towards science seem to be more favorable in laboratory classes in which students perceived greater Student Cohesiveness, Integration, Rule Clarity, Open-Endedness, and a better Material Environment. It is necessary to note that the highest correlation exists between participants' attitudes and Integration subscale scores. This suggests students who perceive greater integration between laboratory activities and theory classes have more positive attitudes toward science. These findings were consistent with other study results reporting a strong link between student outcomes and their perceptions of the science laboratory learning environment (Gidding & Waldrip, 1996; Lee & Fraser, 2001; Wong & Fraser, 1994).

Relative to student achievement, positive correlations were found with the nature of science laboratory learning environment (except Open-endedness). This finding suggests that perceiving laboratory lessons more open-ended, does not necessarily increase students' achievement in science. It is necessary to note that in the present study academic achievement of students were measured by using their previous science grade. This may not reflect their actual achievement in science. Therefore, we considered it as one of the limitations of this study.

Gender in this study was not found to be a significant factor that differentiates students' perceptions of laboratory learning environment. This finding did not support previous research that demonstrated significant gender difference in science laboratory learning environments (Henderson, Fisher, & Fraser, 2000; Quek et al., 2002; Wong & Fraser, 1994; Wong & Fraser, 1997). The finding implied that the science laboratory created such a learning environment that both girls and boys could find equal opportunities to practice with the materials, engage in discussions, and interact with their peers and teachers. Further research, however, is necessary to clarify this finding.

Investigating the Turkish school students' perceptions of the science laboratory learning environment and their association with both attitudinal and cognitive outcomes may provide science teachers, science educators, and curriculum developers with valuable information to make the laboratory more effective place to work in. The findings of the current study suggest that in order to enhance seventh grade students' achievement in science and their attitude toward science, laboratory learning environments can provide students with adequate material, positive social interaction, clear rules to follow, and integration with theory components of the course.

IMPLICATIONS FORM STUDIES OF STUDENT PERCEPTIONS OF SCIENCE LABORATORY ENVIRONMENTS

The results of the present study, as well as the number of other studies reporting investigations on laboratory learning environments, have important implications for science classroom practice. The results provide science teachers with instruments and information about process on aspects of the laboratory learning environment that, if altered, could lead to increases in students' attitudinal and achievement gain. That is, science teachers have the resources, through research as teachers (action research), to make improvements in their laboratories to help promote more positive attitudes toward science and in turn help create a more supportive environment for teaching and learning. Future research needs to examine other variables that could be involved with students' learning environments in their laboratory classrooms, such as grade level, age, school types, student motivation, and teaching approach. A qualitative study could be conducted to obtain a more detailed picture of the science laboratory learning environment and the complex way variables interact.

REFERENCES

Aldridge, J. M., Fraser, B. J. & Huang, T. C. I. (1999). Investigating classroom environments in Taiwan and Australia with multiple research methods. *Journal of Educational Research, 93,* 48-57.

Arisoy, N., Cakiroglu, J., & Sungur, S. (2007, April). *A canonical analysis of learning environment perceptions and motivational beliefs.* Paper presented at the annual meeting of the American Educational Research Association. Chicago, IL.

Cakiroglu, J., Tekkaya, C., & Rakici, N. (2007). Learning environments, teacher interpersonal behavior, and students' attitudinal outcomes in science classrooms: A Turkish context. In K. Mutua & C. S. Sunal (Eds.), *The enterprise of education research on education in Africa, the Caribbean and the Middle East* (pp. 291-311). Greenwich CT: Information Age

Cakiroglu, J., Telli S., & Cakiroglu, E. (2003, April). *Turkish high school students' perceptions of learning environment in biology classrooms and their attitudes toward biology.* Paper presented at the annual meeting of American Educational Research Association, Chicago.

Fraser, B. J. (1990). *Individualised classroom environment questionnaire.* Melbourne: Australian Council for Educational Research.

Fraser, B. J. (1994). Research on classroom and school climate. In D. Gabel (Ed.), *Handbook of research on science teaching and learning* (pp. 493-541). New York: Macmillan.

Fraser, B. J. (1998). Classroom environment instruments: Development, validity and applications. *Learning Environments Research, 1,* 7-33.

Fraser, B. J. (2003). Learning environments. In B. J. Fraser & K. G. Tobin (Eds), *International Handbook of Science Education* (pp. 527-623), Dordrecht: Kluwer.

Fraser, B. J., Giddings, G. J., & McRobbie, C. J. (1995). Evolution, validation and application of a personal form of an instrument for assessing science laboratory classroom environments. *Journal of Research in Science Teaching, 32,* 399-422.

Fraser, B. J., & Treagust, D. F. (1986). Validity and use of an instrument for assessing psychosocial environment in higher education. *Higher Education, 15,* 37-57.

Geban, O., Ertepinar, H., Yilmaz, G., Altin, A., & Sahbaz, F. (1994, September). *The effect of computer assisted instruction on students' science achievement and attitude.* Paper presented in National Science Education Symposium, Izmir, Turkey.

Giddings, G., & Waldrip, B. G. (1996, April). *A comparison of science laboratory classrooms in Asia, Australia, South Pacific and USA: An international study.* Paper presented at the annual meeting of the American Educational Research Association, New York.

Goh, S. C., & Fraser, B. J. (1998). Interpersonal teacher behavior, classroom environment and student outcomes in primary mathematics in Singapore. *Learning Environments Research, 1,* 199–229.

Green, S. B., Salkind, N. J, & Akey, T. M. (2000). *Using SPSS for windows: Analyzing and understanding data.* Upper Saddle River, NJ: Prentice-Hall.

Henderson, D., Fisher, D., & Fraser, B. (2000). Interpersonal behavior, laboratory learning environments, and student outcomes in senior biology classes. *Journal of Research in Science Teaching, 37,* 26–43.

Hofstein, A., & Cohen, I. (1996). The laboratory environment of high school students in chemistry and biology laboratories. *Research in Science and Technological Education, 14,* 103-117.

Hofstein, A., & Lunetta, V. N. (1982). The role of the laboratory in science teaching: Neglected aspects of research. *Review of Educational Research, 52,* 201–217.

Hofstein, A., Nahum, T. L., & Shore, R. (2001). Assessment of the learning environment of inquiry-type laboratories in high school chemistry. *Learning Environments Research, 4,* 193-207.

Lazarowitz, R., & Tamir, P. (1994). Research on the use of laboratory instruction in science. In D. Gabel (Ed.), *Handbook of research on science teaching and learning* (pp. 94-128). New York: Macmillan.

Lee, S. S. U., & Fraser, B .J. (2001, March). *High School Science Classroom Learning Environment in Korea.* Paper presented in the National Association for Research in Science Teaching, St. Louis, MO.

Moos, R. H., & Trickett, E. J. (1974). *Classroom environment scale manual* (1st ed.). Palo Alto, CA: Consulting Psychologist Press.

Ousbourne, J., Simon, S., & Collins, S. (2003). Attitudes towards science: A review of the literature and its implications. *International Journal of Science Education, 25,* 1049-1079.

Quek C. L., Wong, A. F. L., & Fraser, B. J. (2002). Gender differences in the perceptions of chemistry laboratory classroom environments. *Queensland Journal of Educational Research, 18,* 164-182.

Schreiber, J. B., Stage, F. K., King, J., Nora, A., & Barlow, E. A. (2006). Reporting structural equation modeling and confirmatory factor analysis results: A review. *The Journal of Educational Research, 6,* 323-337.

Taylor, P. C., Fraser, B. J. & White, L. R. (1994, April). *CLES: An instrument for monitoring the development of constructivist learning environments.* Paper presented at the annual meeting of the American Educational Research Association, New Orleans, LA.

Telli, S., Cakiroglu, J., & den Brok, P. (2006). Turkish secondary education students' perceptions of their classroom learning environment and their attitude towards biology. In D. L. Fisher & M. S. Khine (Eds.), *Contemporary approaches to research on learning environments: world views* (pp. 517-542). Singapore: World Scientific.

Tsai, C. C. (2003). Taiwanese science students' and teachers' perceptions of the laboratory learning environments: Exploring epistemological gaps. *International Journal of Science Education, 7,* 847-860.

Walberg, H. J. & Anderson, G. J. (1968). Classroom climate and individual learning. *Journal of Educational Psychology, 59,* 414-419.

Waldrip, B. G., & Fisher, D. L. (2000, April). *City and country students' perceptions of teacher-student interpersonal behavior and classroom learning environments.* Paper presented at the annual meeting of American Educational Research Association, New Orleans, LA.

Wong, A., & Fraser, B. J. (1994, April). *Science laboratory classrooms environments and students attitudes in chemistry classes in Singapore.* Paper presented at the annual meeting of the American Educational Research Association, New York.

Wubbels, T., & Levy, J. (1993). *Do you know what you look like?: Interpersonal relationships in education.* London: The Falmer Press.

CHAPTER 6

INVESTIGATING PROCESS-BASED WRITING IN CHEMISTRY LABORATORIES

Andrea Gay

This study investigated the introduction of the decision/explanation/observation/inference (DEOI) laboratory notebook and report writing methodology in two sections of an introductory organic chemistry laboratory course using a qualitative, action research methodology. The DEOI method required students to explicitly describe the purpose of procedural steps and the meanings of observations. Students generally preferred the DEOI method to journal-style laboratory reports used in previous courses and claimed that the method aided their understanding of experiments. The DEOI method was used by students in two distinct manners: recursively writing and revising as intended and contemplation only after experiment completion. Recursive use may have been influenced by teaching assistant attitudes towards the method and seemed to also engender a sense of preparedness. Students' engagement with the DEOI method highlights the promise of constructivist writing in teaching laboratories and the need to focus on the best means to engage students in understanding rather than only using traditional discipline writing formats.

The Impact of the Laboratory and Technology on Learning and Teaching Science K-16, pp. 135–166
135

INTRODUCTION

The "Writing Across the Curriculum" movement asserts that writing should be incorporated in all disciplines, not just English, to promote better writing skills, communication skills, and content learning (Connolly & Vilardi, 1989; Kovac & Sherwood, 2001). The Writing Across the Curriculum movement has begun to influence chemistry teaching as its benefits become more apparent (Shires, 1991), and many colleges have integrated more writing into their chemistry courses. This chapter examines how writing is incorporated into chemistry laboratory courses and how constructivist science writing in particular has been beneficial to student learning. One constructivist science writing method, the decision/explanation/observation/inference (DEOI) method, is examined in detail. The DEOI method was developed as a means of introducing constructivist writing to existing, traditional laboratory courses. This chapter investigates student and teaching assistant responses to using the DEOI method in an organic chemistry laboratory course and how the method seemed to impact student connections between theory and practice in the laboratory. The benefits and challenges of utilizing the DEOI method are discussed with an emphasis on the impact of teaching assistants implementing the method with either a process or product orientation. The implications of the DEOI method for the Writing Across the Curriculum movement in college science laboratories are then outlined.

WRITING IN CHEMISTRY

Writing in College Chemistry Laboratories

Laboratory reports are an integral part of science courses and offer an opportunity to foster science writing within the chemistry curriculum. A number of different approaches have been taken by chemistry professors seeking to focus on student writing within the laboratory. In a general chemistry course, Tilstra (2001) assigned her students to examine a different section of an article from *The Journal of the American Chemical Society* (*JACS*) each week in order to learn how to write a laboratory report in *JACS* format. From student interviews, Tilstra ascertained the method helped students to appreciate the importance of lab reports by working with professional science writing and additionally she noted the assignment improved students' writing skills. Alternately, to help students appreciate the variety of styles of writing that are used in chemistry, Olmsted (1984) had his upper level chemistry students write their lab reports in a different style each week. For example, 1 week the students wrote their results as a popular science article and the next week as a journal style abstract. Olmsted found it difficult to assess if student writing improved since no writing style was repeated during the semester. Koprowski (1997) had his students critique three journal articles after writing their first major lab report to better understand the format of science writing. For their second major lab report the students used a double-blind peer review system to get feedback on their writing before revising. Students reported that they felt the peer review helped make them better science writers. Each of the above writing assignments sought to improve student writing by teaching students greater facility in using the standard writing formats of professional chemists. However, writing-to-learn methodologies need not emphasize the writing styles of the discipline; indeed alternative writing structures may be better at facilitating conceptual understanding of laboratory experiments.

Writing-to-learn methodologies can be coded into three broad categories: modernist, constructivist, and poststructuralist (Prain & Hand, 1996). Modernist approaches center on writing and reading the technical language of science and incorporate more traditional laboratory report and laboratory notebook formats. Constructivist writing-to-learn methods focus on students translating science concepts into their own words and use student writing as a means to engage student thinking. Poststructuralist methods, which are rarely used in chemistry laboratories, are concerned with students understanding broad social and ethnographic implications of science and the use of science. The categories of modernist, constructivist, and poststructrualist can be used by educators as lenses to understand the intended purposes and methodologies of laboratory writing assignments and exercises (Prain, 2006).

Most chemistry laboratory manuals advocate writing that approximates the writing of chemists and thus tend to promote a modernist approach to writing. The beginning of almost every published laboratory textbook outlines how to arrange a laboratory notebook to include procedures and observations. Lab reports are to be organized in a professional journal-based format, including abstract, introduction, methodology, results, discussion, and conclusions. Alternately, some approximation of the journal style format maybe advocated, such as only including results and discussion or using a premade data table, which is easier to grade and less taxing on the students.

Writing journal articles is a primary public writing activity for professional chemists. For scientists, writing is an integral means of constructing meanings and clarifying ideas in both public writing and personal writing, such as bench experimental notebooks (Yore, Bisanz, & Hand, 2003). This is the case even if scientists have difficulty articulating the function of writing in constructing knowledge (Yore, Hand, & Florence, 2004), or indeed even if their stated purposes in writing are simply to transmit knowledge (Yore, Hand, & Prain, 2002). While journal style laboratory reports facilitate meaningful learning for scientists, they are often not productive for students because "the transformation of ideas" that "occurs during the write-react-revise phases of argumentation" for science experts is absent for novice writers who generally "do not emphasize the write-revise sequence in their writing process" (Yore et al., 2003, p. 713). However, students have had more success with traditional laboratory reports when they are explicitly guided through the write-revise-sequence, such as with the computerized LabWrite program, which directs students though each section of their laboratory report (Ferzli, Wiebe, & Carter, 2004)

Journal style laboratory reports may engage thinking when they are coupled with inquiry-based experiments because science inquiry itself fosters conceptual thinking. However, traditional organic chemistry laboratory courses often include only descriptive or verification experiments and rarely incorporate inquiry-based laboratory activities. Using journal style lab reports with descriptive or verification experiments can hinder meaningful learning. Keys (1999) explained that in verification experiments "learners quickly realize that they must somehow generate, copy, or paraphrase the knowledge claim that is desired by the teacher" and thus laboratory report writing "can easily become a rote activity" (p. 125). When using journal style reports, many students can get caught up in writing information in different sections in the manner the professor requires without understanding why (Chinn & Hilgers, 2000) or can become focused on formatting and templates instead of content (Keys, 1999; Pickering & Goldstein, 1977).

Knowledge Transformation in Chemistry Writing

Writing can be used as a tool by students in the process of learning and not just as an artifact or product of prior learning. Bereiter and Scardamalia (1987) make the distinction between "knowledge telling" and "knowledge transformation" writing tasks. "Knowledge telling" occurs when the learner writes ideas they have previously formulated, in other words, the products of their thought. Knowledge transformation occurs when the learner has to alternate between content and rhetoric cognitive domains. The learner must consider a topic and then consider how to articulate that idea into writing. The resultant writing then serves as new input into the cognitive domain.

The strengths of the Writing Across the Curriculum movement are most apparent in constructivist writing for process, or knowledge transformation, rather than product, or knowledge telling (Connolly & Vilardi, 1989; Kovac & Sherwood, 2001). Innovative writing assignments, such as mini-essays, science journalism, and laboratory journals, have been successfully used in college chemistry courses. Wilson (1994) had his students write mini-essays describing reaction mechanisms. Wilson found that the exercise helped students to make connections and it was easier for him to spot student misconceptions. Using science journalism has increased student understanding and engagement in introductory chemistry courses, whether by having students find chemical mistakes in the popular press and correct them (Venable, 1998) or writing articles for a broad audience on chemistry experiments (Hallowell & Holland, 1998). Hermann (1994) had her organic students keep a traditional laboratory notebook and a laboratory journal. In the journal the students kept track of their successes, failures, and approaches for future work. The journal writing helped them to foster reflective thinking. As indicated by the aforementioned studies, constructivist writing in chemistry has the potential to increase student conceptual understanding.

Laboratory reports are an important genre for students to consider and communicate work in the laboratory and are well established within chemistry curricula (Keys, 1999) providing a convenient platform for introducing constructivist writing into courses. Recent research has started to investigate how laboratory reports can be transformed to focus on student scientific thinking rather than assuming student reasoning will be enhanced though using traditional genre formats. For instance the Model, Observe, Reflect, Explain (MORE) thinking framework serves to scaffold student thinking within a inquiry based laboratory format (Tien, Rickey, & Stacy, 1999). The framework includes writing, reviewing, and revising as they make models of their systems more dynamic. However, utilizing the MORE framework has required large-scale curricular reform.

The (SWH) has been used very effectively with a wide age range of students, from middle school to college (Hand, Wallace, & Yang, 2004; Hohenshell & Hand, 2006; Keys, Hand, Prain, & Collins, 1999; Poock, Burke, Greenbowe, & Hand, 2004; Rudd, Greenbowe, & Hand, 2001/ 2002; Rudd, Greenbowe, Hand, & Legg, 2001). The SWH consists of six components:

(a) *Beginning ideas and questions*—What are my questions about this experiment?
(b) *Tests and procedures*—What will I do to help answer my questions?
(c) *Observations*—What did I see when I completed my tests and procedures?
(d) *Claims*—What can I claim?

(e) *Evidence*—What evidence do I have for my claims? How do I know? Why am I making these claims?

(f) *Reading and discussion*—How do my ideas compare with others? (Rudd et al., 2001/2002, p. 231)

The SWH is similar to traditional lab report formats, although it has a different emphasis. The traditional discussion section in which students refer to classroom theory, numerical results, and error sources is replaced by "claims" and "evidence" sections, which require students to relate results to their own conceptual understandings. The SWH seeks to promote a more inquiry-based approach even if the laboratory manual is not significantly altered. However, the SWH still requires clear experimental questions.

Traditional synthetic organic chemistry experiments often have no clear experimental questions beyond, "Can molecule A be made by procedure B?" The answer to which is always "Yes." Procedures are explicitly detailed with few opportunities for students to make choices about how to conduct the experiment. Hopefully, college faculty will become more familiar with science education theory and university structures will further support reform in college courses and laboratories. This would allow increased use of inquiry and problem-based instruction and utilization of laboratory writing methods such as MORE and SWH. Until that time, steps towards increasing laboratory reforms should incorporate more research on writing-to-learn laboratory report formats that can be used with either inquiry experiments or chemistry experiments as they are now.

INVESTIGATING THE USE OF THE DEOI IN CHEMISTRY LABORATORY INSTRUCTION

Development of the DEOI Writing Method

In response to the need for constructivist writing methods that can be used with the traditional synthetic organic chemistry experiments currently found in many universities, I created the DEOI laboratory writing methodology. The DEOI method was developed as writing-to-learn laboratory notebook and report methodology, with an emphasis on revision and self-explanation. The method was also designed to help students forge connections between chemical theory and their practices in the laboratory. Students using the new process writing methodology divided their laboratory notebooks into four columns decision/explanation/observation/inference expanded from the traditional procedures/observations.

The expansion was devised to make students explicit in their reasoning and to help them make connections between chemical theory and manipulations in the lab. The decision column was intended for listing procedural steps. In the explanation column were listed the rationales for experimental design, explanations of why particular protocols were used, and explanations of what was occurring on a molecular level. In the observation column, the actual observables in the lab such as smells and colors were listed, as well as measurements and spectral data. The inference column was for interpreting observations, making judgments, and planning future courses of action. The students were supposed to fill in the decision and explanation columns of the DEOI schema in their laboratory notebook before coming to class and add to the schema during the laboratory experiment. Then the students revised their schemas based on their experiences in the laboratory and turned in the revised DEOI schema as part of the lab report. An example of a student DEOI schema can be found in the Appendix.

Students were instructed to write the decision column and their initial ideas for the explanation column in their notebooks before coming to the laboratory. Writing before the lab was intended to make students consider their work prior to conducting an experiment since many students tend not to plan ahead and can, hence, be unprepared to work in the laboratory (Heppert, Ellis, Robinson, Wolfer, & Mason, 2002). Having students explicitly write an explanation for each decision, or procedural step, required students to make their reasoning explicit as to why they were using certain techniques and reagents, even if all of the procedural steps were categorically given to them. Furthermore, since oral self-explanation protocols can impact science problem solving and reading comprehension (Bielaczyc, Pirolli, & Brown, 1995; Chi, De Leeuw, Chiu, & Lavancher, 1994; Wong, Lawson, & Keeves, 2002), written self-explanation protocols such as the explanation and inference columns may also aid in student understanding. During the laboratory session students filled in the observation and inference columns and had the opportunity to make changes to the explanation column once they had seen the actual reactions and performed the techniques. The inference column was separated from observation to help students differentiate the ideas of claims versus evidence, which many students initially find difficult (Abd-El-Khalick, Lederman, Bell, & Schwartz, 2002; Gunstone & White, 1990). After conducting the experiment, students revised their DEOI schemas based on their experiences in the lab and turned in the revised DEOI schema as part of their lab report, which included analysis of spectra, reaction mechanisms, an overall discussion of the experiment, and end of unit questions. The students schemas were designed to be working documents to test ideas, and the students were encouraged to discuss the intricacies of the experiment with their laboratory teacher assistants (TAs), classmates, or lab partners. In the DEOI method the revised schema was not just a product to be graded, but also a template for thinking and rethinking about the processes and concepts of the experiment.

Prain and Hand (1996) described three hallmarks of constructivist writing tasks that are effective at promoting learning through writing. Writing tasks should "require students to elaborate, reprocess concepts and central ideas, hypothesize, interpret, synthesize and persuade, and hence develop higher-order reasoning skills" (Prain & Hand, 1996, p. 618). Second, students should have to grapple with unfamiliar as well as familiar content in the writing. Finally, the writing task should be constructed to enable students to self-monitor to see if they are on task. The DEOI method fulfilled these requirements. In completing the explanation and inference columns students should have reviewed central concepts, made sense of procedural steps and observational data, and defended conclusions. When students engaged with new techniques and new reactions they encountered theory with which they were unfamiliar. The ease or difficulty students experienced in attempting to complete the explanation and inference columns could also indicate to students where their conceptual understanding was strong and where it was still weak.

Research Population, Methodology, and Data Sources

The DEOI method was introduced into two sections of introductory organic chemistry laboratory at a major research university in New York City. Of the 17 students in the study, 13 were postbaccalaureate (postbacs) students who had completed undergraduate degrees and were enrolled in the course to fulfill premedical (premed) requirements in order to apply to medical school. The postbacs' college degrees were in a wide variety of disciplines in the humanities, natural sciences, and social sciences. The other four students in the course were undergraduates, three of whom were also premeds. Nine of the students had worked as research assistants in either psychology or biology laboratories giving them some experience with research science; however, none of the students had worked previously in a chemistry research laboratory. Ethnically the student population was quite diverse with eight Caucasian students, four students of East Asian descent, two students of South East Asian descent, one Hispanic student, and two students who self-identified as bi-racial, specifically African American/White and White/Hispanic.

There were TAs for each of the two sections: Mark, a first year graduate student who had worked in industry for 2 years, and Chris, a first year graduate student who had come to graduate school directly from completing his undergraduate degree (All names are pseudonyms). Nine students were in Mark's section, five of whom were postbacs, and eight students were in Chris's section, all of whom were postbacs. The chemistry department insisted that because of the TAs' many commitments as graduate students, no training could be implemented beyond that in which all TAs in the department participate. Furthermore, the TA meetings were required to focus on preparing for the logistics of the upcoming experiment and quiz rather than pedagogical instruction for the TAs. In response to these constraints, Mark and Chris were given a TA version of the laboratory manual that sought to elucidate the educational rationale for the DEOI method as well as providing teaching tips and answers to end of unit questions.

Student and TA use of and responses to the DEOI method were investigated using an action research methodology. Action research can be fruitful means for examining classroom interventions when the researcher is both innovator and a participant observer. The process of identifying a problem in the classroom, planning a course of action to address the problem, and then researching the effects of that action constitute the basic structure of action research. Ideally this process is circulative, so that the results of research guide the restructuring, and the new plans are in turn researched in the pursuit of best practice (Arhar, Holly, & Kasten, 2001; Creswell, 2002). I identified a problem when I was an organic chemistry laboratory instructor and observed students' difficulties using traditional journal style lab reports and connecting theory and practice in the laboratory. Developing the DEOI method was a course of action seeking to ameliorate student learning difficulties. During implementation I acted as a participant observer researching the effects of the DEOI method and also served as secondary resource TA for the students.

Implementation of the DEOI method was investigated by employing the three Es of action research: experiencing, enquiring, and examining (Creswell, 2002). While the TAs knew that the researcher was also the developer of the DEOI method, the students were not made aware of this unless they specifically asked. As a participant observer, I wrote field notes of experiences in the laboratory, discussions with students, and conversations with TAs. In semistructured interviews, students and TAs were asked about their opinions of and interactions with the writing method. Interviews lasting from 20-40 minutes were conducted with 14 of the 17 students and both TAs. Care was taken in interviews to elicit negative as well as positive responses. Each TA was asked to select one high achieving student, one low achieving student, and two average students. Revised DEOI schemas, laboratory reports, and quizzes of these eight students were examined for student conceptual understanding and TA grading. A demographic survey was administered to the students at the end of the semester. The department does not administer exams in laboratory courses. Data was not gathered from the lecture course because while some students were taking lecture and laboratory concurrently, others had completed the lecture course a year or more prior to the laboratory course.

Analysis was cyclical and initial coding began during data collection to guide development of interview protocols. After completion of the course, the complete interview transcripts, field notes, and surveys were read to provide a view of the data as a unified whole before formally beginning the process of coding and segmenting it into thematic pieces. Formal coding began by organizing data by the overarching research questions, centered on student and teaching assistant perceptions of the DEOI method and evidence of students connecting theory and practice. Data were then further subdivided as either positive, neutral, or negative responses. These data groups were then reread for emergent themes, and the themes re-examined for additional detailed coding. A highly branched coding scheme developed from this process. Positive themes were compared with the themes that emerged from neutral and negative groupings to determine mixed responses and to ensure that challenges as well as successes of the method were represented in the final analysis. The revised DEOI schemas and the field notes were then re-read to see if the themes students and TAs discussed were evident in their work and actions in order to triangulate data and to determine if new undiscussed themes became apparent.

Results of Implementation

Student Use of the DEOI Method in Both Sections

The aim of the DEOI method is to provide students with a tool to positively impact their learning in the laboratory. Most students in both sections cited that they preferred the DEOI method to journal style laboratory reports, which they had used in other courses. The students reported the DEOI method was more efficient, more focused on their own thinking, and helped scaffold their thinking about the experiment. Furthermore, there was evidence in both Mark's and Chris's sections of students using the DEOI method to better understand the experimental procedures and observations. However, students did experience some challenges in learning to use the method especially at the beginning of the semester.

Efficiency and Scaffolding of Student Thinking

Students appreciated the DEOI method and believed it aided their learning. Having taken previous laboratory courses most of the students were familiar with traditional laboratory writing formats. When comparing the DEOI method to traditional reports they had used in the past, most students favored the DEOI method. Their preference for the DEOI method is particularly compelling given that they had to write a DEOI schema for all eight experiments in the course, while students in the non-treatment sections of the course only had to complete two journal style laboratory reports and write a short summary of results at the end of the lab period for the other six experiments.

The students found the DEOI method more efficient and more focused on thinking than the formalisms of writing style. Kumar said that the DEOI methodology "made things a lot easier, made it a little more practical than, um, it's a lot easier than writing up a whole report, definitely." A few of the students who had worked in biology research laboratories stated that journal style laboratory reports seemed inappropriate for routine classroom experiments with known outcomes and found the DEOI method more appropriate for introductory chemistry laboratory. Patrick noted that the more efficient format was also more focused on student thinking and stated, "I did think that the shorter lab write up were much more focused and helped me focus anyway on the important aspects." Echoing Patrick's sentiments, Joseph explained his preference:

> [In the DEOI method] as long as you understood the experiment like why you did all the steps is the only really important thing. So I mean I think that was good, especially compared to that [journal style reports]. I mean with that thinking isn't even part of the equation. It's just like write copious discussion like whatever, even if you have nothing important to say, something fairly trivial you just have to like just keep writing and writing, so, yeah, I think it was good.

Most students preferred using the DEOI method to journal style lab reports that they had used in prior courses with only one student, Pershant, desiring more experience with formal lab reports since he plans to do research in the future. However, Pershant did note that experience with the DEOI method still provided "the same line of reasoning" as a formal report.

Some of the students stated that the method helped scaffold and organize their ideas while in the laboratory. The DEOI columns became a framework for understanding organic chemistry experiments. For Corina, the DEOI method "gave you like a structure of what was happening" to aid in thinking about and preparing for the experiment. Patrick expanded on this notion, "It was nice to see exactly where, what step did what to produce what results and it was all… it was nice to have it all ordered in column form and read across and everything makes sense." The format of the DEOI schema, according to Patrick helped him easily connect how certain procedures lead to specific observations. While many students appreciated how the DEOI method helped scaffold their thinking, a few students did state that at times the method felt overly structured especially toward the end of the semester.

Student Understanding of Procedures and Observations

Students appreciated the emphasis on thinking and felt that the method aided in their understanding of procedures and the experiment. The explanation and inference columns most notably stood out to students as hallmarks of the method. Annik praised the explanation column,

> I feel like the explanation part was a good one because that's where you really learn to understand why you need to be doing certain steps.... Understanding why you're going to be doing things certain way, why you're going to be using this versus that.

Andy concisely stated that for him the benefit of the Explanation and Inference columns was "just like transferring from the cookbook to like understanding what each step means." Other students were appreciative that the explanation and inference columns forced them to consider their reasonings. Helena explained:

> Just I think it's very good to try and actually explain every step because it makes you understand what you are doing. So it makes you, like instead of like mechanistically adding stuff and being clueless about why, it makes you think ... it definitely forced you because otherwise I don't think I would. (laughs)

Students felt that the DEOI method helped them better understand the experiments because it provided a platform for understanding and required them to use that platform. Mark, one of the TAs, said that in his section the previous semester, which did not use the DEOI method, students tended to ask, "Why are we doing this?" while his students using the DEOI method were more likely to ask, "Are we doing this because of this?"

The DEOI method was constructed to help foster student understanding about the reasons for using certain procedures and the meaning of observations. Some of the students' revised DEOI schemas highlight understanding of procedures and observations that could easily be overlooked in a traditional laboratory writing format that focuses primarily on the final product. For instance, to understand the procedures, Joseph described the reason for the intricate construction of a column chromatography apparatus:

> The cotton prevents sand from washing out the bottom of the column. The bottom layer of sand prevented very fine silica from passing through

column. Silica acts as stationary phase. Top layer of sand prevents top surface of silica from being deformed by solvent added on top. If surface was made uneven, it's possible that two separate bands would wash into collecting test tube simultaneously. This is why it's important to keep the column perfectly straight up and down. (Experiment 3-Explanation column: Exp. 3-E)

Similarly, Helena showed depth of reasoning and attention to her experimental work as she sought to understand the meaning of anomalous observations during a Wittig reaction,

> Observation: When the mixture is transferred into the addition funnel, it yellows (the mixture). Inference: The appearance of yellow tint might be due to the NaOH [sodium hydroxide] mixture reacting with benzyltriphenylphosphonium chloride remaining on the neck of the flask to form yellow ylide. (Exp. 7- O&I) (See Appendix for Helena's complete Exp. 7 DEOI schema.)

There was evidence in the revised DEOI schemas of some students demonstrating depth of reasoning and knowledge of the meaning of their procedures and observations. It is impossible to unequivocally state that this depth was due to using the DEOI method, especially since higher-level reasoning was not apparent in all students' revised schemas. However the students' statements that the DEOI method specifically helped them focus on understanding the reasons for procedures and the meanings of observations suggest that the evidence of conceptual complexity in the revised schemas is related to use of the DEOI method. Students approved of the DEOI method citing that it was more efficient and more focused on students' own thinking than journal style laboratory reports, that the method helped structure their thinking about the experiments, and that completing the Explanation and Inference columns aided their understanding of concepts.

Challenges Using the Method

Students' responses and laboratory reports also illuminated aspects of the method that need to be reexamined for future implementations. The students had some difficulties with understanding how to use the DEOI method, especially at the beginning of the semester, which is to be expected whenever students are asked to try a new methodology. As is common for students in the laboratory (Gunstone & White, 1990), for Eric "it wasn't very clear what [is] the difference between observation or inference." Karen discussed her initial difficulties saying,

It was just a little confusing in like how to write it up each section.... I was confused in the beginning in terms of like do I need to put a thing in the, you know, Inference column for everything.

The most common mistake of this sort was writing information that should have been in explanation column, which should contain the reasons for procedural steps, in the inference column, which is the space for noting the meaning of observations. For instance, Kumar included in his revised DEOI schema, "Observation: Used glass rod to scrape the inner walls of flask. Inference: Glass rod helps in overall rate, same amt., faster time" (Exp. 4-O&I). It may be that as students watched themselves perform procedures, especially those not detailed in the laboratory manual, the procedures became observations to them. The students who experienced the most confusion of which column was appropriate for different types of information tended to have the lowest grades in other aspects of the course such as quizzes and problem sets. While this confusion generally abated by the end of term, more concrete examples and more explicit TA feedback on graded work should be given to students in the future.

It is impossible to alleviate all student confusion associated with completing the DEOI columns since some student challenges simply arose from students grappling with difficult concepts and pushing the boundaries of their understanding. Corina noted that she sometimes had difficulty giving reasons for procedures in the explanation column, "just because I just didn't really know how to explain it." Confusion was at times an indicator that students were engaged in thinking about the content.

Effects of Differing Implementation of the DEOI Method

The DEOI method was well received by most students who felt that the requirement to explain procedures and observations helped them understand the experiment and structure their thinking. Students in both sections of the course perceived these benefits despite the fact the DEOI method was used by students in two very different ways. These two ways of using the DEOI method were elucidated though observations, discussions with TAs, and student interviews. The first way was process based and comprised of completing the decision and explanation columns before class, adding to the schema during class, and revising afterwards. When used as intended, the DEOI method has the potential to help students connect theory and practice *while in the laboratory setting* and to repeatedly engage students in the write-revise sequence. The second way was to utilize the DEOI in a modernist format by copying the decision column before class and contemplating the explanations and reasonings entirely at home after the experiment was complete. Used in this manner, students can blindly conduct reactions in the laboratory and only come to make meaning of their actions *after finishing the experiment*. Interviews with students indicated that, while not unilaterally, Mark's students tended to try to use the method the first way, recursively as intended. Chris's section tended to use the DEOI method in the second way in an unintended modernist manner.

A number of factors were investigated to understand the root of the differences in implementation in the two sections. One possibility was that by chance students in Mark's section had more of an orientation to using writing to clarify thinking rather than using writing to demonstrate proficiency. However, no correlation was found between students' actual use of the method and their declared belief of the purpose of academic writing (field notes and interviews). Mark was more enthusiastic about teaching and often tried to show his students how equipment or experiments that the students used in the teaching laboratory were related to his personal research work (field notes). Chris was less concerned about teaching because he felt that most of his students were only taking the course to fulfill premed requirements and that it did not matter if he brought outside chemistry material into the classroom (interview). It is possible that the difference in the two sections was due to the fact that Mark enjoyed teaching and devoted more of his energies to teaching than did Chris; that Mark had progressed further in his development as a TA (Nyquist, Abbott, & Wulff, 1989; Nyquist & Wulff, 1996). However interviews and observations of students' strong attachment to Chris and approval of Chris's teaching methods seems in conflict with this idea.

Both Mark and Chris espoused fairly traditional views of what chemistry topics should be taught. For instance, when discussing possible changes to a Grignard reaction experiment both made suggestions to have the experiment better mimic what they used in their research laboratories rather than considering changes based on learning outcomes (field notes). Both felt that the purpose of lab was primarily to give a hands-on counterpoint to what students learned in the lecture class. For instance Chris said that in the laboratory "you actually do a type of reaction that you have learned in class so it's something that's hands on instead of just looking in a book." However, they differed in their belief of the importance of process versus product in the laboratory. Mark believed in having students understand each step of the procedure and encouraged constructivist use of the DEOI method as a process based writing methodology. Chris conveyed to his students a focus on results and products in the laboratory encouraging more modernist use of the methodology. The TAs orientation to either process or product can be seen in their reactions to the DEOI method and student use of the method and in their grading of student work. TAs with an orientation to process may be able to facilitate greater benefits for students using process based writing methodologies.

Product Orientated Use of the DEOI Method

Chris preferred using journal style laboratory reports and briefer summaries of results and discussion. He felt that these writing methods should be used instead of the DEOI method in future courses. Chris acknowledged that his preference might be based in familiarity:

> Maybe I was just more biased to previous or the old way because that's just the way that I was taught. (laughs) I guess, I guess that's maybe why I didn't like this way so much. Just because that's what I'm used to.

Furthermore, Chris understood the purpose of the method saying, "I think using the old way they're more focused on how much you get and if it's good, bad," whereas with the DEOI "new method it's not really focused on like how much you get or if you get any as long as you understand why you got what you got." Yet, he felt that difference may be ineffectual:

> I don't know if a ... student actually realizes that it is different. Because I think when a student goes to lab I think they're automatically like "I need to get," you know, "a certain amount [compound yield] to get a certain grade." So I think they already think that before they even start anything.

Chris believed that his students were already and unchangingly in the mindset of products and results associated with journal style laboratory reports.

Chris also felt that the DEOI method was too difficult to grade. He rarely checked prelab writing. As a process based system there were not set responses for the DEOI columns that all students should have, unlike a product-orientated system with a definite focus on outcomes such as percent yield and melting point comparisons. He preferred the traditional lab "where you get a sheet that tells you what to look for" in the lab reports. Chris found the lack of set answers in the DEOI method difficult and to some extent found the method confusing. He elaborated, "I didn't know exactly what was supposed to be there. And it was, I guess, that was why it was hard to grade... because it wasn't cut and dry." His grading reflected that Chris, at least to some extent, continued to grade in the manner of a product-based system. For instance, the only comments he made on the collected samples of student revised DEOI schemas for Experiment 4 were comments such as "% yield" and "Next time draw structure and arrows from H's to peaks" on Jennifer's NMR spectra. All of Chris's comments were about results, such as yield, melting point, and spectral analysis, which are the focus of traditional lab report methodologies. He made no comments concerning the process of experimentation and observations in the laboratory. Chris's grading difficulties may also have stemmed from the fact that he was "still confused on what the difference between observation and inferences" and had trouble grading what he himself found to be somewhat difficult.

Chris was used to modernist writing methods that focus on results, and he preferred grading results based writing. He believed that his students were already in the mindset of products and results associated with traditional laboratory reports. It is probable that Chris's attitude about the virtues of the DEOI method, which manifested in his estimate of whether the method should be used again and in his grading affected how his students used the method. Xenia, one of Chris's students, understood the purpose of the DEOI method:

> I think that if you do it [prelab writing] thoroughly you probably get more out of the lab. I'm sure. I mean you know what to look for. You know exactly what's going on at every step instead of kind of figuring it out in hindsight what happened.

Yet she hesitantly discussed how she felt Chris's actions affected her own:

> At least in our lab, you know, it just was a very (pause) checking our prelabs that wasn't like this vigilant process, so we tended to kind of work on what we had to turn in after lab more that what we were supposed to before the lab. Is it bad that I'm saying this? (laughs)

If Chris had placed more importance on the benefits of DEOI method, his students may have as well.

In the DEOI method students are supposed to recursively revise their DEOI schemas before, during, and after the experiment. The DEOI method is less effective, when it is not implemented as intended. When the emphasis in the laboratory is on the product of the experiment rather than the process, laboratory work can easily become blindly following a recipe. For instance, Eric and Patrick, who were laboratory partners in Chris's section, were unable to give an answer during lab as to why they were adding potassium hydroxide (KOH) to a ketone during an aldol condensation reaction, which is a key step in the reaction forming the enolate. They did not understand the purpose of adding KOH was because they had both written the decision column in their notebooks before lab, but neither had attempted to write anything in the explanation column. Eric and Patrick simply added KOH because the manual said to add KOH (field notes). However, considering Patrick's lab report it was evident that he came to understand the purpose of base in enolate formation. A primary purpose of the DEOI method is to help students connect theory and practice in the laboratory, but if students are guided to have a product approach and use the method in a modernist manner, the connections between theory and practice are often made only after the fact. This can diminish the importance of laboratory experiences and weaken the connections that are made.

Furthermore, many of Chris's students found prelab writing to be "tedious." Chris's students tended to copy the procedure into the decision column before laboratory and then not attempt to complete the explanation column until after the experiment. When used in this manner, the prelab writing became not only time consuming but also fairly meaningless rather than part of the learning process. Andy said, "to blindly recopy something you could, could like not even be there re-copying it." According to Annik, prelab writing was too time consuming in part "because you had to spend the work after class anyway," and she did not work on understanding the experiment until after class.

Students in Mark's section also found rewriting the procedures to be somewhat tedious, but they tended to have a different perception of the prelaboratory writing in general. The sense of tedium was due in large part to most students directly copying the procedure from the laboratory manual since they had difficulty translating the procedure into their own words as evidenced by examination of student work and interviews. Joseph suggested giving students a spreadsheet with the procedures written in the decision column precisely so students could focus their prelaboratory energies on the explanation column, which would "still force you like to go through the procedure and understand it." Pershant stated, "this requirement that we have to sort of explain the procedure before getting to class was probably the most useful like, you know, aspect of this lab." For students who used the DEOI method recursively as intended, prelaboratory writing was an asset. The following describes how Mark facilitated process based writing for his students and the additional benefits of using the DEOI method which his students accrued.

Process Orientated Use of the DEOI Method

Mark was enthusiastic about the DEOI method. He felt that the focus on understanding procedure fit well with his own philosophies of teaching and preferred the DEOI method over other writing schemes typically used in the department. Mark explained his preference:

> The goal of the course is so for them to understand like what they're doing in the lab and so, you know, that's just a more formal way of assessing whether they actually understood it rather than just relying on technique points…. So, you know, other than technique points, which like I said no

one really does much with that, there's no way, there was really no way to evaluate their actual understanding of what they were physically doing in the lab in the other format. Whereas in [the DEOI method], this format there is definitely, that's like a major component of lab, which it should be. I mean it should be a like a major component of the class: why, like explain it, understanding what we're doing. I mean why else be in here, might as well not go to lab! So, definitely it was, definitely better.

Mark placed a major emphasis on students understanding the procedures and the process of completing the experiments and appreciated the process writing because it emphasized these goals. Mark did assert that in the future he would like to see the DEOI method used in conjunction with a formal paper. He believed that "there are plenty of people that, you know, have a really, really hard time writing scientific writing" and to address that he "would have made them do one [formal report] at least." However, he felt that the possibility of combining the DEOI method with traditional methods would work specifically because his students had become so adept at using the DEOI method. He said, "they all got into a groove … [and] got more efficient at it but once they were at that point that would have been a fine time to throw a paper at them." Mark, as a practicing chemist, would have liked to see students have opportunities for formal writing, but he thought that those opportunities could and should be in conjunction with the DEOI method.

Mark's approach to grading reflected his ideas of a focus on process. While he was not stringent in what was included and gave students full credit as long as they had written something, he checked student prelab writing each week (field notes). While he discussed that the DEOI method put more pressure on TAs because students have to know more to be able to fill out the columns, he felt it was good because the revised schemas could show where students were having problems with the material (field notes). His comments on student revised DEOI schemas often corrected problems students encountered and asked them to focus on important aspects of procedure. For instance, Joseph guessed the observed "white foam is aspirin crystallizing out of solution. Reacting vigorously because acid is strong and solution is ice cold thus reducing solubility." (Exp. 2-I) Mark responded, "Actually foam is CO_2 [carbon dioxide] bubbling out as $NaHCO_3$ [sodium bicarbonate] is neutralized." He would also praise exemplary work such as Helena's insight that the carbon dioxide spike seen in her infrared spectra was due to the fact that "CO_2 is soluble in KBr [potassium bromide], even more than it is in air" (Exp. 4-I).

Mark believed in the necessity of students understanding processes in the lab and felt that the DEOI process writing methodology helped students achieve this goal. Mark detailed his thoughts:

> Like even though that wasn't part of the [regular] course [last semester] I was still always trying to get them to understand why they were doing everything even though they didn't really need to know to get a good grade in the class…. [The DEOI method] changed my approach this semester in the sense that it made it easier for me to do that. I didn't have to like do it all on my own because they kind of did it on their own to start so they all have some idea of what they were doing…. And they would have like an explanation ready even though a lot of times it was wrong but at least they have something to give me that I could that way I could at least know what they were thinking and know how best to like get them not to think in the wrong way or get them, or reinforce them when they were thinking in the right ways.

His students echoed his perceptions of their learning in the laboratory.

Mark's students who used the DEOI recursively cited that they felt more prepared and more efficient in the laboratory. Corina said that, "it forces you to know ahead before coming to class, so you can plan ahead what going to happen in the lab and you're more prepared." The DEOI method helped prepare them to understand what they were doing in the laboratory while they were in the laboratory. Pershant explained what he felt was the most useful attribute of the course:

> I found that like especially with like the Explanations and like working through the procedure before you got to the lab, it helped me, you know, know what I was doing most of the time when I was doing the experiment.

For Karen being prepared translated to being able to be more efficient in the lab. She explained:

> if you set up the [DEOI] schema, there are certain things that you know that you can do ahead of time and just have ready. But there are things that you're like "Oh, ok, we should wait … because otherwise it's going to evaporate or it's going to like degrade a little bit." I think in that sense it was like really good for us to be more time efficient.

Kumar noted that when he worked on his explanation column before the laboratory it helped him "understand sort of the reason behind the methodology" and that "it becomes easier just to draw inference from your observations if you've done that." The DEOI method encouraged some students to ask more questions in the laboratory because as Joseph stated, "I know what questions to ask, so I can write the thing up." Chris did not mention if the DEOI method helped the students in his section be more prepared, more efficient, or ask pertinent questions in the laboratory. Only one of his students, Katrina who earned the highest grade in Chris's section, cited the benefit that "it makes you think about what you're going to be doing." The underlying theme of Mark and his students' perceptions is that by using the DEOI method recursively, writing the Explanation column before class, they felt more prepared and efficient because they were able to connect theory and practice *while conducting their experiments*. These benefits of the DEOI method are only seen with correct implementation of the method and highlight the need for focusing on TA education.

Conclusions

The Importance of TA Education and Fidelity in Teaching with Innovations

As evidenced by the differing use of the DEOI method in Chris and Mark's section, TA beliefs impact the implementation of laboratory writing reforms. Poocke et al. (2004) found similar effects with the implementation of the Science Writing Heuristic (SWH) in a general chemistry course. While all of the students benefited from using the SWH, the best student outcomes were observed in sections where the TA implemented the writing method with high fidelity. Unfortunately departmental constraints disallowed additional TA training for Mark and Chris. However, for students to fully utilize the DEOI writing-to-learn methodology in the future and garner its benefits to the fullest degree more high quality TA education is needed.

Luft, Kurdziel, Roehrig, and Turner (2004) wished to examine why university TA training programs and teaching experience are ineffective. They compared traditional TA education to "growing a garden without water." The TAs felt that their instruction was not practical enough, there was little mentorship since professors' attentions primarily rested on research, and ideas about science education tended to be experiential and intuitive rather than based on scholarship. Mark and Chris's only official training consisted of attending a departmental TA training session at the beginning of their graduate school career. Research on educational beliefs of general chemistry TAs in the department indicate that their training experiences were similar to those described by Luft et al. (Gay, 2001). This is not to condemn the practices of the chemistry department because exemplary TA educational measures can be difficult to introduce since they require a large investment of time and a substantial shift in departmental ethos. However, it is not enough to change how students engage with content, attention must be paid to how TAs teach the content. This is the case whether looking to increase interactive teaching methods in recitation sections (Landis et al., 1998; Nurrenbern, Mickiewicz, & Francisco, 1999), using inquiry-based laboratory experiments (Heppert et al., 2002), or introducing writing-to-learn laboratory report methodologies (Rudd et al., 2001). Effective TA training programs are extensive and ongoing, provide mentoring, and place a priority on student learning (Ambers, 2002; Druger, 1997; McComas & Cox-Petersen, 1999; Nurrenbern et al., 1999).

Writing-to-Learn in the Laboratory

Despite differing implementation of the DEOI method by TAs, both sections of students evidenced benefits from using the DEOI method. When comparing the method to journal style laboratory report styles which they had used in prior courses, many students stated that the DEOI method was less stressful, more efficient, and provided a welcome emphasis on thinking rather than stylistic formalisms. While the SWH, which is structured similarly to a traditional laboratory report but with a much greater emphasis on students formulating questions, claims, and evidence, is substantially different from the DEOI method, college students' comparisons between the SWH and traditional laboratory report formats were very similar to students' comparisons between the DEOI method and traditional laboratory reports. General chemistry students using the SWH reported that they felt they learned more in less time and that the heuristic helped them focus on thinking unlike previous laboratory courses (Rudd et al., 2001; Rudd et al., 2001/2002).

Many students felt that the using the DEOI method helped them understand their procedures and observations in the laboratory. They stated that the method encouraged them to think, especially when completing the Explanation column. The students reported the DEOI method allowed them to consider their actions, and some students appreciated that it required them to consider their actions because they would not necessarily have been inclined to do so on their own. The TAs also noted that the DEOI method was beneficial in helping students understand the concepts and procedures in the laboratory. Again this response is similar to general chemistry students using the SWH who felt that the heuristic encouraged their internal motivation and that explanations of their work in the laboratory aided their understanding (Rudd et al., 2001) and to introductory biology students using the online laboratory writing tutor LabWrite who felt their laboratory reports made them more observant in the laboratory and assisted their conceptual understanding (Ferzli et al., 2004).

The perceived benefits of explaining their procedures and observations in the laboratory may stem from the conceptual understanding benefits attributed to self-explanation. Students who perceived that the DEOI method aided in their understanding might be accruing the similar benefits as found in oral self-explanation protocols. Chi et al. (1994) hypothesized that the cognitive benefits of oral self-explanation may manifest because it is a construction activity where students are required to form new declarative and procedural knowledge and formulate rules of when to use that knowledge because it encourages integration of new information with the students' prior knowledge, and because it allows students to continuously update their mental models as they compare them to the text or problem at hand. These characteristics are very similar to the hallmarks of constructivist writing tasks that emphasize students formulating their own ideas, the interplay between new information and prior knowledge and ideas, and metacognition (Prain & Hand, 1996).

SUMMARY AND IMPLICATIONS OF WRITING-TO-LEARN IN THE LABORATORY

Both the SWH and the DEOI method were formulated as constructivist writing methodologies emphasizing student understanding rather than traditional modernist writing formats which focus on writing as used by scientists (Prain, 2006; Prain & Hand, 1996). The similarities in responses to these two differently structured laboratory report writing methods may indicate that students are favorably responding to constructivist laboratory report writing irrespective of the details of the method used. Constructivist based laboratory report writing methodologies, such as the DEOI method, have the potential to help students connect theory and practice in the laboratory and to deepen their understanding of chemical concepts. The DEOI method can be used with typical verification experiments as are found in many university chemistry departments. Utilizing the DEOI method is a viable means of introducing constructivist writing and introducing educational reforms in the laboratory.

Continued Research

Further research on the DEOI method will elucidate how the method should be tailored for differing student populations and inform how students are connecting theory and practice through using constructivist writing methodologies. Action research is recursive with results informing changes that are in turn investigated (Arhar et al., 2001; Bogdan & Biklen, 1998; Creswell, 2002). This study should be viewed as an initial investigation of the DEOI method. The challenges that arose from use of the method should be addressed before a reiteration of implementation. Particularly students need to be initially given more information about and better examples of the DEOI method in the laboratory manual, and they need TAs to provide more feedback as they use the DEOI method.

TAs need to have more formalized training which focuses on how to implement and grade the DEOI method, as well as probing discussion of TAs' ideas about teaching and learning and an introduction to the science education literature. Future research could incorporate using the revised DEOI method with a larger number of sections and comparing qualitatively and quantitatively learning outcomes of students using the DEOI method and students using traditional laboratory reports in control sections. Special attention would need to be paid to constructing quantitative measures because the types of connections between theory and practice that the DEOI method seeks to foster are not typically tested. The DEOI method was initially developed to be used with traditional organic chemistry laboratory experiments, but investigating the use of the method with inquiry based experiments would shed light on how the DEOI method can be incorporated into larger laboratory reform efforts.

The DEOI method is an example of a process based, constructivist laboratory report writing methodology. Continued research of constructivist writing methodologies is critical to further inform how student science literacy can be enhanced (Prain, 2006) Organic chemistry is a course ripe for these studies specifically because students often find the course very challenging and laboratory procedures are typically very prescribed. Future research should also examine the effectiveness of constructivist writing methods with differing student populations both with regard to content area, such as non-majors chemistry, and student demographics, for instance at an urban, historically minority university. Continued work with the DEOI method as an example of process-based writing will inform the research community about how students are using writing as a means of connecting theory and practice, as well as inform practitioners on the best means of implementing constructivist writing in their own classrooms to impact the learning of their students.

APPENDIX

Revised DEOI Schema for Experiment 7:
Synthesis of Stilbene Using the Witting Reaction
by Helena (Undergraduate Premedical Student in Mark's Section)

Decision	Explanation	Observation	Inference
I. Formation of Stilbene			
1. Set up a 150ml three-necked flask with a condenser, and an addition funnel. Add to the flask 2g of benzyltriphenylphosphonium chloride, 20ml of methylene chloride, and a stir bar.		7.810g of benzyltriphenylphosphonium chloride added – granules, white and somewhat transparent. Benzaldehyde is a clear liquid, density 1.05g/ml – added 2ml, which is actually a bit more than 2g – approximately 2.1g. Methylene chloride is a clear volatile liquid.	
2. Stir vigorously with a magnetic stirrer. Add 25ml of 50%by weight aqueous NaOH via the addition funnel. The base should be added 5ml at a time.	Stir to evenly distribute the reagents. Addition of base removes HCl from benzyltriphenylphosphonium chloride to form a phosphorous ylide. The ylide, once formed, reacts with benzaldehyde to form stilbene as well as the side product Triphenylphosphine oxide. Base is added slowly for safety reasons since the reaction is exothermic.	The mixture turns yellow when base is added. As more base is added, the mixture darkens in color and appears almost orange. No noticeable increase in temperature. Base is very dense, which makes it difficult to add it in large amounts via the addition funnel – add drop wise.	Temperature did not increase rapidly because the base was added rather slowly, drop by drop. The color of the mixture is due to the formation of brightly dark colored ylides, the colors of which range from yellow to orange to red. As time goes by, more and more ylides are formed, the mixture appears to be darker.

Appendix continues on next page.

Appendix Continued

Decision	Explanation	Observation	Inference
3. The temperature will rise slightly. Stir the solution for 30min. the solution should turn cloudy and yellow after about 5min.	Temperature rises because the reaction is exothermic.	The mixture reaches a point at which it is the darkest in color (orange), after which it lightens in color and the process seems to be reversed. The solution goes from being dark orange to yellow to white/transparent. Solution turns a bit cloudy and 2 layers are visible – bottom layer is made up of white crystal-like particles, the top layer is yellow then white transparent liquid	When the solution becomes the brightest/darkest in color – it contains the maximum amount of unreacted ylide. As ylide is formed, it reacts with benzaldehyde to form white/colorless stilbene. Therefore, as the concentration of ylide in the solution decreases, the color of the solution changes from orange to yellow to finally white/transparent. At the point when the solution is colorless, the reaction is over because there is no more ylide present. The solids present in the solution could be the following: stilbene, byproducts, or precipitated NaCl (it could precipitate since the aqueous solution used is so highly concentrated)

II. Isolation and Purification of Stilbene

Decision	Explanation	Observation	Inference
1. Using a separatory funnel, separate the organic layer from the aqueous layer. Wash the organic layer 2-3 times with 20ml portions of water. Wash the organic layer with a saturated solution of sodium bisulfite and then again with water. Repeat the sodium bisulfite and water washing until the pH is neutral.	Sodium bisulfite neutralizes the base NaOH. When testing for pH, test the aqueous layer since litmus paper works the best with aqueous solutions. The pH of the two layers, organic and aqueous is identical.	When the mixture is transferred into the addition funnel, it yellows (the mixture). After testing, it was determined that the organic layer is the one on top. Used about 100ml of water and 30ml of sodium bisulfite.	The appearance of yellow tint might be due to the NaOH in the mixture reacting with benzyltriphenylphosphonium chloride remaining on the neck of the flask to form yellow ylide.

2. Dry the organic layer over anhydrous magnesium sulfate. Filter the solution into an Erlenmeyer flask and evaporate to dryness.	Drying removes all water, while evaporation removes the rest of organic solvent, thus isolating the product.	Added very little of anhydrous magnesium sulfate. Not many clumps formed. Crystals formed are somewhat flaky and clumpy.	There must have been little H2O present, thus a small amount of the drying agent was necessary.
3. Add 20ml of hexane to the crude product stir well and then filter. Repeat with 2 more 20ml portions of hexane. Combine the hexane extracts.	Triphenylphosphine oxide is fairly insoluble in hexane while stilbene is readily soluble in hexane.	Some crystals were rather large, and were not dissolved in hexane. Some of the crystals remained in the flask stuck to its walls.	Large crystals could have possibly contained stilbene in them, which was not accessible for the hexane to dissolve. Thus, some of the product was lost and the yield will decrease.
4. Pack a chromatography column with silica gel and moisten the column with hexane. Add the 10ml of concentrated product hexane solution. Elute the column in a single fraction using 30ml of hexane-methylene chloride solution. Use TLC to monitor the column to make sure no other side products begin to elute through the column. Collect the Chromatography fraction in a tared flask.	Column chromatography is used to selectively collect the product. Since stilbene is soluble in hexane, it will elute first.	Used about 100ml of the hexane methylene system. The TLC testing did not confirm decrease in concentration of stilbene elution. The column was stopped short. Flask weight is 55.085g	Since the concentration of an organic product, either stilbene or Triphenylphosphine did not decrease, the elution of the column was either incomplete or overdone. Thus, either too little stilbene was eluted, or all the stilbene and some side product were eluted. In the first case, the effect may be that the ration of cis to trans product would be incorrect if the two types of product elute at different times (have different polarity). In the second case, the ratios of the product will be correct, but the product will be contaminated with side product.

Appendix continues on next page.

Appendix Continued

Decision	Explanation	Observation	Inference
5. Evaporate the solution to dryness. Weigh the product and calculate percent yield.		The solution did not evaporate to dryness after 20 min of Rotovap; the flask contained some crystals and some liquid in it. After several days, the product flask still has some yellowish liquid in it intermixed with grayish fleck-like crystals. Final product weight $56.011g-55.085g = .926g$	Not all of the solvent evaporated. If the liquid has stilbene dissolved in it, the yield of stilbene will decrease. If, however, the liquid is one in which stilbene is not soluble, yield will not be affected. The cis isomer is a liquid at room temperature. Thus, the crystals are probably the trans isomer, while the liquid is cis.

III. Analysis

Decision	Explanation	Observation	Inference
1. Analyze the product using GC. Dissolve a small amount of product in deuterated chloroform. Inject 5mmL of the solution. Note the GC settings. Inject sample of authentic cis-stilbene and trans-stilbene to determine identity of each peak in the product mixture.	Ration of peak heights helps to determine the percentage composition of isomers in the product.	Oven temp 200C Flow rate 45 Injection temp 250C Cis isomer shows up before the trans isomer. Mixture of one pipette tip of product to one pipette of solvent was too dilute – GC did not show two product peaks. More concentrated mixture was used – 1:10 of product: solvent – successfully. Solvent peak is the greatest in size. GC shows the ratio of cis: trans isomers to be about 3:1	
2. Analyze mixture using TLC. Use a dilute solution since stilbene isomers have similar Rf values and it will be difficult to see the separation of the solution is too concentrated. 50/50 hexane/methylene chloride mixture will work as a developing solvent. Visualize with UV light.		Both the pure trans-stilbene and the product were spotted. The spots were intense, though the product did not separate into two clear spots. Rather, the spots were overlapping. However, they were separated enough to signify presence of two different compounds.	Since not a very good separation of the two isomers was attained, the hexane/methylene chloride system used was not suitable. It was either too polar or not polar enough. The TLC confirms that the product was pure since only two overlapping spots show up.

| 3. Take HNMR spectrum in CdCl3. | HNMR spectrum can be used to determine the ratio of cis-stilbene to trans-stilbene in the product mixture. Then this ratio can be compared to the ratio determined by GC. Use deuterated chloroform because deuterium does not show up in the same place on the HNMR spectrum when hydrogen splitting usually show up. | Cis isomer shows up at lower ppm values. A large peak for chloroform shows up. The ratio determined by HNMR is approximately 2:1 cis: trans. | The solvent "CDCl3" was not pure. It was about 99.8% deuterated. The hydrogenated chloroform was present in enough amounts to appear on the spectrum. |

Note: Revised DEOI schema copied directly from Helena's work including grammatical errors and non-conventional writing of chemical formulas. Her complete laboratory report also included analysis of spectra, reaction mechanisms, a summary discussion of the experiment, and end of unit questions.

REFERENCES

Abd-El-Khalick, F., Lederman, N. G., Bell, R. L., & Schwartz, R. S. (2002). Views of nature of science questionnaire (VNOS): Toward valid and meaningful assessment of learners' conceptions of nature of science. *Journal of Research in Science Teaching, 39*(6), 497-521.

Ambers, R. K. R. (2002). Learning about teaching: A graduate student's perspective. *Journal of College Science Teaching, 31*(5), 327-330.

Arhar, J. M., Holly, M. L., & Kasten, W. C. (2001). *Action research for teachers: Traveling the yellow brick road*. Upper Saddle River, NJ: Merrill Prentice Hall.

Bereiter, C., & Scardamalia, M. (1987). *The psychology of written composition*. Hillsdale, NJ: Erlbaum.

Bielaczyc, K., Pirolli, P. L., & Brown, A. L. (1995). Training in self-explanation and self regulation strategies: Investigating the effects of knowledge acquisition activities on problem solving. *Cognition and Instruction, 13*(2), 221-252.

Bogdan, R. C., & Biklen, S. K. (1998). *Qualitative research for education: An Introduction to theory and methods* (3rd ed.). Boston: Allyn & Bacon.

Chi, M. T. H., De Leeuw, N., Chiu, M.-H., & Lavancher, C. (1994). Eliciting self-explanations improves understanding. *Cognitive Science, 18*(3), 439-477.

Chinn, P. W. U., & Hilgers, T. L. (2000). From corrector to collaborator: The range of instructor roles in writing-base natural and applied science classes. *Journal of Research in Science Teaching, 37*(1), 3-25.

Connolly, P., & Vilardi, T. (Eds.). (1989). *Writing to learn mathematics and science*. New York: Teachers College Press.

Creswell, J. W. (2002). *Educational research: Planning, conducting, and evaluating quantitative and qualitative research*. Upper Saddle River, NJ: Merrill Prentice Hall.

Druger, M. (1997). Preparing the next generation of college science teachers: Offering pedagogical training to graduate teaching assistants as part of the college reform agenda. *Journal of College Science Teaching, 26*(6), 424-427.

Ferzli, M., Wiebe, E., & Carter, M. (2004, April). *College students' perceptions about lab reports: A response to in-depth instruction*. Paper presented at the annual meeting of the National Association for Research in Science Teaching, Vancouver, BC.

Gay, A. (2001, October). *Graduate teaching assistants' views of group learning in a general chemistry laboratory*. Paper presented at the Association for the Education of Teachers of Science Northeast Regional Conference (AETS-NE), University of Syracuse, New York.

Gunstone, R. F., & White, R. T. (1990). Promoting conceptual change in the laboratory. In E. Hegarty-Hazel (Ed.), *The student laboratory and the science curriculum* (pp. 291-299). New York: Routledge.

Hallowell, C., & Holland, M. J. (1998). Journalism as a path to scientific literacy. *Journal of College Science Teaching, 28*(1), 29-32.

Hand, B. M., Wallace, C. W., & Yang, E.-M. (2004). Using a Science Writing Heuristic to enhance learning outcomes from laboratory activities in seventh-grade science: quantitative and qualitative aspects. *International Journal of Science Education, 26*(2), 131-149.

Heppert, J., Ellis, J., Robinson, J., Wolfer, A., & Mason, S. (2002). Problem solving in the chemistry laboratory: A pilot project to reform science teaching and learning. *Journal of College Science Teaching, 31*(5), 322-326.

Hermann, C. K. F. (1994). Teaching qualitative organic chemistry as a writing intensive course. *Journal of Chemical Education, 71*(10), 861-862.

Hohenshell, L. M., & Hand, B. M. (2006). Writing-to-learn strategies in secondary school cell biology: A mixed methods study. *International Journal of Science Education, 28*(2-3), 261-289.

Keys, C. W. (1999). Revitalizing instruction in scientific genres: Connecting knowledge production with writing to learn in science. *Science Education, 83*(2), 115-130.

Keys, C. W., Hand, B. M., Prain, V., & Collins, S. (1999). Using the science writing heuristic as a tool for learning from laboratory investigations in secondary science. *Journal of Research in Science Teaching, 10*(36), 1065-1084.

Koprowski, J. L. (1997). Sharpening the craft of scientific writing: A peer review strategy to improve student writing. *Journal of College Science Teaching, 27*(2), 133-135.

Kovac, J., & Sherwood, D. W. (2001). *Writing across the chemistry curriculum: An instructor's handbook*. Upper Saddle River, NJ: Prentice Hall.

Landis, C. R., Peace Jr. , G. E., Scharberg, M. A., Branz, S., Spencer, J. N., Ricci, R. W., et al. (1998). The New Traditions Consortium: Shifting from a faculty-centered paradigm to a student-centered paradigm. *Journal of Chemical Education, 75*(6), 741-744.

Luft, J. A., Kurdziel, J. P., Roehrig, G. H., & Turner, J. (2004). Growing a garden without water: Graduate teaching assistants in introductory science laboratories at a doctoral/research university. *Journal of Research in Science Teaching*, *41*(3), 211-233.

McComas, W. F., & Cox-Petersen, A. M. (1999). Enhancing undergraduate science instruction: The G-step approach. *Journal of College Science Teaching*, *29*(2), 120-125.

Nurrenbern, S. C., Mickiewicz, J. A., & Francisco, J. S. (1999). The impact of continuous instructional development on graduate and undergraduate students. *Journal of Chemical Education*, *76*(1), 114-119.

Nyquist, J. D., Abbott, R. D., & Wulff, D. H. (Eds.). (1989). *Teaching assistant training in the 1990s* (Vol. 39). San Francisco: Jossey-Bass.

Nyquist, J. D., & Wulff, D. H. (1996). *Working effectively with graduate assistants*. Thousand Oaks, CA: Sage.

Olmsted, J., III. (1984). Teaching varied technical writing styles in the upper division laboratory. *Journal of Chemical Education*, *61*(9), 798-800.

Pickering, M., & Goldstein, S. L. (1977). The educational efficiency of lab reports. *Journal of Chemical Education*, *54*(5), 315-317.

Poock, J., Burke, K., Greenbowe, T. J., & Hand, B. (2004, April). *Evaluating the effectiveness of implementing inquiry and the Science Writing Heuristic in the general chemistry laboratory.* Paper presented at the annual meeting of the National Association for Research in Science Teaching, Vancouver, BC.

Prain, V. (2006). Learning from writing in secondary science: Some theoretical and practical implications. *International Journal of Science Education*, *28*(2-3), 179-201.

Prain, V., & Hand, B. (1996). Writing for learning in secondary science: Rethinking practices. *Teaching and Teacher Education*, *12*(6), 609-626.

Rudd, J. A., Greenbowe, T. J., & Hand, B. (2002). Recrafting the general chemistry laboratory report. *Journal of College Science Teaching*, *31*(4), 230-234. (Original work published 2001)

Rudd, J. A., Greenbowe, T. J., Hand, B. M., & Legg, M. J. (2001). Using the science writing heuristic to move toward an inquiry-based laboratory curriculum: An example from physical equilibrium. *Journal of Chemical Education*, *78*(12), 1680-1686.

Shires, N. P. (1991). Teaching writing in college chemistry: A practical bibliography 1980-1990. *Journal of Chemical Education*, *68*(6), 494-495.

Tien, L. T., Rickey, D., & Stacy, A. M. (1999). The MORE thinking frame: Guiding students' thinking in the laboratory. *Journal of College Science Teaching*, *28*(5), 318-324.

Tilstra, L. (2001). Using journal articles to teach writing skills for laboratory reports in general chemistry. *Journal of Chemical Education*, *78*(6), 762-764.

Venable, T. L. (1998). Errors as teaching tools: The mass media mistake. *Journal of College Science Teaching*, *28*(1), 33-37.

Wilson, J. W. (1994). Writing to learn in an organic chemistry course. *Journal of Chemical Education*, *71*(12), 1019-1020.

Wong, R. M. F., Lawson, M. J., & Keeves, J. (2002). The effects of self-explanation training on students' problem solving in high-school mathematics. *Learning and Instruction, 12*(2), 233-262.

Yore, L. D., Bisanz, G. L., & Hand, B. (2003). Examining the literacy component of science literacy: 25 years of language arts and science research. *International Journal of Science Education, 25*(6), 689-725.

Yore, L. D., Hand, B., & Florence, M. K. (2004). Scientists views of science, models of writing, and science writing practices. *Journal of Research in Science Teaching, 41*(4), 338-369.

Yore, L. D., Hand, B. M., & Prain, V. (2002). Scientists as writers. *Science Education, 86*(5), 672-692.

CHAPTER 7

THE EFFECT OF THE VEE HEURISTIC ON STUDENTS' MEANINGFUL LEARNING IN PHYSICS LABORATORIES

Tarek Daoud and Saouma BouJaoude

The purpose of this study was to investigate the effectiveness of using the Vee Heuristic to organize student experiences in the Grade 10 physics science laboratory. The Vee Heuristic type of laboratory was compared to a traditional verification type of laboratory on students' attitudes toward physics and physics laboratories and their achievement in physics. Participants in this study were secondary school students enrolled in two sections of a laboratory-based Grade 10 physics course (a treatment and a control group). Five instruments were used in the study: Two attitude questionnaires, a physics achievement test designed to measure meaningful learning in physics, a Vee Heuristic Scoring Key, and an Individual Interviews protocol. The treatment was administered over 12 weeks. Results indicated that using the Vee Heuristic in physics laboratories in contrast to verification laboratories did not enhance students' meaningful learning of the target physics concepts; however the Vee Heuristic approach had a significant and favorable effect on students' attitude toward physics and the physics laboratory.[1]

The Impact of the Laboratory and Technology on Learning and Teaching Science K-16, pp. 167–200

INTRODUCTION

The science laboratory has been considered one of the main vehicles for enhancing science learning not only as a means of demonstration but also as the heart of the science learning process (Hofstein & Lunetta, 1982, 2004; Nakhleh, 1994; Tamir, 1989; Trowbridge & Bybee, 1990). According to Ausubel (1963) laboratory activities help students to appreciate science and the methods used in scientific discovery. In order to reach this goal, the learning process in the laboratory must be based on the constructivist view of learning in general and inquiry in particular. Using inquiry provides students with the opportunity to engage in the process of investigation (Abd-El-Khalick et al., 2004; Carillo, Lee, & Rickey, 2005; Hofstein & Lunetta, 1982, 2004; Roth & Roychoudhury, 1993). Research on the effects of laboratory activities shows that involving students in authentic inquiry helps them to construct their own understandings of science concepts (Roehric, Lulie, & Edwards, 2001; Roth & Roychoudhury, 1993), construct knowledge schemes and solve problems (German, Haskins, & Auls, 1996), practice science as scientists (Okebukola & Ogunniyi, 1984), and develop positive attitudes toward science (Lehrer, Schauble, & Petrosino, 2001) and critical thinking and decision-making skills (Abd-El-Khalick et al., 2004).

Okebukola and Ogunniyi (1984) showed that secondary level science students working in cooperative groups in laboratories achieved higher grades and used more science process skills than students in individualistic or competitive groups. Similarly, Rubin and Tamir (1988) showed that high school biology students using application-investigative laboratories achieved higher grades than students using verification type laboratories. Tamir (1977) compared verification-type laboratory and the inquiry-based laboratories and concluded that in verification-type laboratory instruction, the teacher identifies the problem to be investigated, relates the investigation to previous work, conducts demonstrations, and gives explicit guidelines that students have to follow. Tamir adds that in an inquiry laboratory, the teacher asks the students to formulate the problems, relate the investigation to previous work, state the purpose of the investigation, identify the problem, predict the results, identify the procedures, and perform the investigation. In addition to promoting inquiry, the science laboratory can help students to develop knowledge schemes (Nakhleh & Krajcik, 1993), solve problems (Hofstein & Lunetta, 1982), practice science as scientists (Hofstein & Lunetta, 1982; Okebukola & Ogunniyi, 1984), construct meaningful scientific knowledge (Roth & Roychoudhury, 1993), and acquire cognitive skills such as critical thinking, applying, synthesizing, decision making, and creativity (Lebowitz, 1998; Weaver, 1998).

ASSESSING STUDENT LEARNING IN LABORATORY INSTRUCTION

Despite the suggestion that inquiry laboratories are more efficient than verification-type laboratories, students working in the laboratory are often engaged in the mechanical performance of simple tasks with little interest for developing understanding about the purpose and logic behind such activities (German, Haskins, & Auls, 1996). Laboratory activities, as they are presently practiced, lead students to face difficulties in relating classroom concepts with laboratory experiences (Nakhleh, 1994; Roehric, Lulie, & Edwards, 2001). Novak and Gowin (1984) observed that students are usually confused about the goals of laboratory activities. Moreover, they are not sure what regularities in events or objects to observe or what relationships between concepts are significant. Novak and Gowin continue by suggesting that students "proceed blindly to make records or manipulate apparatus with little purpose and little consequent enrichment of their understanding of the relationships they are observing and manipulating" (pp. 47-48). Similarly, Hofstein and Lunetta (2004) maintained that the science laboratory is not used appropriately to help students acquire concepts or develop inquiry skills. According to Hofstein and Lunetta, a mismatch may exist between teachers' philosophies and their practices and some of them do not seem to be convinced that the laboratory can help students develop meaningful knowledge. In addition, teachers are not trained to use non traditional, constructivist, methods in the laboratory. Nakhleh (1994) reported that students face three difficulties when doing laboratory work: lack of understanding of the basic concepts underlying laboratory activities, inability to relate observations to theoretical knowledge, and inability to organize their observations so that irrelevant details are filtered out. Heinzen-Fry and Novak (1990) asserted that educators have to move students from reliance on rote to meaningful learning. Effective physics instruction should encourage conceptual understanding. Carillo, Lee, and Rickey (2005), report that in addition to "decreasing the recipe laboratory experiments and increasing the use of activities that incorporate student inquiry, students need guidance regarding how to think through the inquiry process" (p. 61).

One of the possible outcomes of engagement in laboratory activities is developing positive attitudes toward science (Hofstein & Lunetta, 2004; Lehrer, Schauble, & Petrosino, 2001; Roth, 1990; Weaver, 1998). Developing favorable attitudes toward science has been identified by educators as one of the important goals of science teaching (Gardner, 1991). This emphasis on developing positive attitudes toward science resulted form research findings indicating that science was the least favored subject for students at the intermediate and secondary levels in many countries (Mason, 1992; Salta & Tzougraki, 2004).

Freedman (1997) and Hofstein and Lunetta (1982, 2004) claimed that attitude toward science is affected by laboratory instruction and that students' attitudes toward and interest in laboratory work were positively correlated. Investigating the effect of using the Vee Heuristic on attitude toward science and toward laboratories is especially important because students' attitudes toward science seem to decline as they progress through school (Butler, 1999; Greenfield, 1997; Simpson & Oliver, 1985; Weiss, 1987). Haussler and Hoffmann (2002) maintained that this decline is especially true in physics for girls. Additionally, research has shown that girls have less positive attitudes towards science compared to boys at the intermediate and secondary levels (Weinburgh, 1995), while other studies showed no differences between girls and boys (Butler, 1999; Salta & Tzougraki, 2004).

Research results on students' attitudes toward science in Lebanon are inconclusive. Raad (1997) and Pharaon (1984) showed that there is no significant difference between males and females regarding the attitudes toward science. On the other hand, Nokari (1998) demonstrated that male students had significantly more positive attitudes toward science than female students, and attitudes toward science decreased with grade level.

USING THE VEE HEURISTIC IN LABORATORY INSTRUCTION

The role of laboratories in science teaching has received increased attention in the past decade with many science educators recommending the use of inquiry to overcome problems associated with laboratory work. Using inquiry in the laboratory serves to shift from teacher- to student-centered teaching and from rote to meaningful learning. One of the meaningful learning techniques that can help students enhance their scientific knowledge, link methodological and conceptual understanding, and develop positive attitudes toward physics is the Vee Heuristic.

The Vee Heuristic is so named because it takes the shape of a letter "V" (Figure 7.1). According to Novak (2002), "The shape serves to give emphasis and distinction to a number of important epistemological elements that are involved in the construction of new knowledge, or new meaning" (p. 549). The Vee Heuristic points to the event being studied from the perspective of a hypothesis or focus question at the top. The left side of the Vee represents the conceptual (thinking) aspects. The right side of the Vee represents the methodological (action or doing) aspects. Focus questions bring these two sides of the Vee into active interplay as students investigate the events or objects relevant to any particular inquiry. To employ the Vee Heuristic requires that students and/or

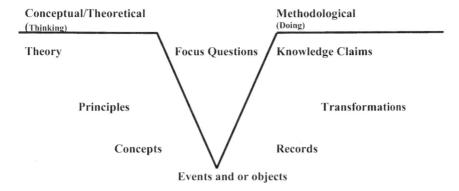

Figure 7.1.   Schema showing the different components of the Vee Heuristic
(Novak & Gowin, 1984).

teachers give explicit consideration to (1) the research question(s), (2) the
event and/or objects to be observed, (3) the relevant concepts and
theories, (4) the data recording and data transformation procedures, and
(5) the knowledge and value claims that derive from the inquiry (Gurley-
Dilger, 1992; Novak & Gowin, 1984; Novak, Gowin, & Johansson, 1983;
Roth & Bowen, 1993; Roth & Roychoudhury, 1993).

A review of the literature did not identify research that examined the
effectiveness of the Vee Heuristic in physics laboratories. Moreover,
except for a number of studies conducted at the college level (e.g., Tessier
& Nakhleh, 1993) the literature review did not identify research studies
that investigated the effects of using the Vee Heuristic on students' attitudes toward science or the science laboratory. However, research has
demonstrated that involving students in scientific inquiry experiences,
comparable to the Vee Heuristic, can potentially affect students' attitudes
toward science positively (Nakhleh, 1994; Nataraj & McManis 2001).
Moreover, since the issue of the possible differential response of males
and females to inquiry activities and to science in general is not resolved
(Von Secker & Lissitz, 1999), it is important to investigate gender differences in response to the Vee Heuristic.

In summary, students do not seem to benefit from laboratory activities
to develop conceptual understanding (Ausubel, 1963; German, Haskins,
& Auls, 1996); relate laboratory and classroom work (Nakhleh, 1994; Roehric, Lulie, & Edwards, 2001); develop inquiry skills (Hofstein & Lunetta,
2004); and understand the relationships between concepts, activities, and
evidence (Novak & Gowin, 1984). In addition middle and high school students, especially girls, seem to have negative attitudes towards science in
general and physics more specifically (Haussler & Hoffmann, 2002).

INVESTIGATING THE USE OF
THE VEE HEURISTIC IN LABORATORY INSTRUCTION

The present study was designed to investigate the effectiveness of the Vee Heuristic in developing students' meaningful learning of science concepts and enhancing their attitudes toward physics when working in the physics laboratory. The study was guided by the following research questions:

1. Does using the Vee Heuristic in physics laboratories in contrast to using verification-type laboratories, enhance students' meaningful learning of the target physics concepts?
2. Does using the Vee Heuristic in guiding laboratory instruction enhance students' attitude toward science and the science laboratory?
3. Are there gender differences in attitude and achievement between students using the Vee Heuristic in physics laboratories and those who use verification-type laboratories?

This study may help to provide empirical support for the theoretical arguments about the use of Vee Heuristic; an unexplored representation tool with the potential to increase students' meaningful learning in science or for improving the use of laboratories in science teaching and physics more specifically.

METHOD

Participants

Participants in this study were 47 Grade 10 (ages 15-16 years) high school students enrolled in a physics course at a K-12 Lebanese school in which the language of science instruction is English. The school is a private, independent, and nonsectarian school. It offers a Lebanese and an American curriculum starting at the Grade 1 level. Students in the school mainly come from middle class families with the majority of parents having university degrees. The school has a policy of keeping class size at around 25 students. Students participating in this study were enrolled in the American curriculum.

The school policy requires that students be assigned randomly into different sections, four class sections in the case of Grade 10 physics. At the beginning of the first semester (2006/2007) one of the sections was randomly assigned to be the experimental group while one of the three other sections was randomly selected as the comparison group. There were 24 students in the experimental group (13 males, 11 females) and 23 students in the control group (14 males and 9 females) (see Table 7.1)

The two groups covered the same subject matter, shared the same academic objectives, and took the same achievement and attitude instruments prior to and following the intervention. The content of the physics unit during which the study took place included potential difference (voltage), current, and resistors (Ohm's Law). Students were expected to accomplish the following objectives: (1) Recognize the units of voltage, current, and resistance; (2) describe the role of cells and batteries; (3) explain the notion of electrical current and potential difference; (4) explain the role of resistors; (5) apply the laws of uniqueness and of addition of current and voltage; (6) apply Ohm's Law; (7) apply the law of resistivity; (8) use voltmeters and ammeters to measure voltage and current; (9) describe the direction of the conventional current; 10) differentiate between series and parallel circuits; 11) differentiate between internal and external circuits; and (12) set-up electrical circuits.

The experimental group used the Vee Heuristic during their laboratory sessions while the control group used the regular laboratory activities. The two groups covered the same subject matter, shared the same academic objectives, and took the same achievement and attitude instruments prior to and following the intervention. The experimental group used the Vee Heuristic to guide their laboratory activities while the control group used the traditional textbook laboratory activities that were highly scripted in guiding student actions. Students in the experimental group started by developing a focus question in a whole class discussion. This was followed by relating the focus question to the physics concepts, principles, and theories (left side of the Vee Heuristic)

Table 7.1.   Intervention and Comparison Groups by Gender, and Nationality

Group	Intervention n (%)	Comparison n (%)
Participants	24 (51.06)	23 (48.93)
Gender		
Male	13 (54.20)	14 (60.90)
Female	11 (45.80)	9 (39.10)
Nationality		
Lebanese	20 (83.30)	20 (87.00)
Palestinian	4 (16.70)	3 (13.00)

they had already covered. Then, they worked in groups to plan the experiment (events or objects), collect and record data, organize data in tables, and use the data to answer the focus question and draw conclusions (right side of the Vee Heuristic). Finally, they related the results and conclusion to the concepts, principles, and theories. Alternatively, students in the control group were provided with the focus question and procedure and worked in groups to implement these procedures and answer the focus question.

The teacher who implemented the intervention was trained to use the Vee Heuristic and was provided with lesson plans for each of the lessons designed for the purposes of this study. Additionally, students who used the Vee Heuristic were trained for 2 weeks on how to construct a Vee Heuristic based on the work of Novak and Gowin (1984) and Brody (1986). The treatment study took 12 weeks at the rate of 3 periods per week. The same teacher taught the experimental and the control groups. In order to ensure the authenticity of treatment, one of the researchers selected certain sessions randomly to observe the experimental and control groups. The observations included writing descriptions of the activities taking place during the sessions

Moreover, one of the researchers used the Vee Heuristic scoring rubric to assess the Vee Heuristics constructed by the experimental group students early during the intervention to ensure that students constructed appropriate Vee Heuristics (see Figure 7.2). Finally, a Vee Heuristic test was used to assess students' knowledge of the components of the Vee Heuristic and the relationships between these components. The Test of Physics Achievement (TPA) was used to measure experimental and control students' achievement at the end of the study, while the Physics Attitude Scale (PAS) and the Physics Laboratory Attitude Questionnaire were administered prior to and following the intervention. Finally, individual interviews were conducted with the experimental group students following the study.

Instruments

Instruments used in this study included the Vee scoring rubric, TPA, PAS as Modified from the Fennema-Sherman Attitude Scale, the Physics Laboratory Attitude Questionnaire, and individual interviews.

Vee Heuristic Scoring Rubric

A rubric was used to assess the Vee Heuristics of the students. This rubric was adapted with modifications from Novak and Gowin's (1984) Vee scoring key. This rubric is divided into five main topics: focus

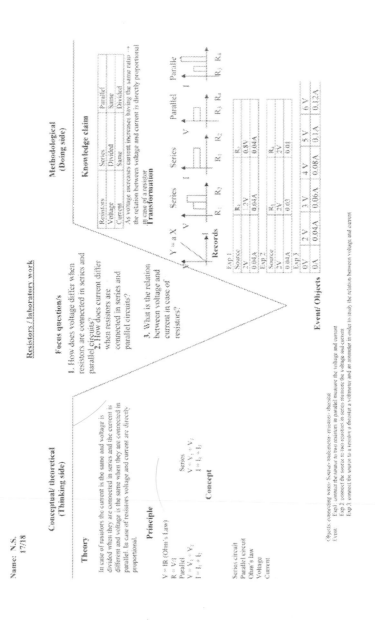

Figure 7.2. A Vee Heuristic produced by a student who received a high score on the scoring rubric.

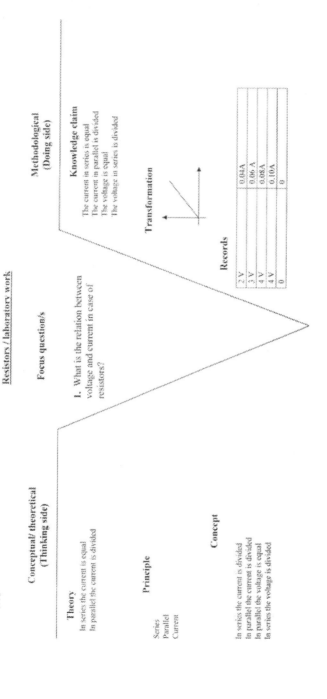

Figure 7.3. A Vee Heuristic produced by a student who received a low score on the scoring rubric.

Table 7.2.   Vee Heuristic Scoring Rubric Showing the Topics and Subtopics Used in the Evaluation of Students' Vee Heuristics

Topic		Item	Score
Focus question	0	No focus question is identified	
	1	A question is identified, but does not focus upon the objects and the major event OR the conceptual side of the Vee	
	2	A focus question is identified; includes concepts, but does not suggest objects or the major event OR the wrong objects and event are identified in relation to the rest of the laboratory exercise.	
	3	A clear focus question is identified; includes concepts to be used and suggests the major event and accompanying objects.	
Objects/Event	0	No objects or event are identified.	
	1	The major event OR objects are identified and are consistent with the focus question, OR an event and objects are identified, but are inconsistent with the focus question.	
	2	The major event with accompanying objects is identified, and is consistent with the focus question.	
	3	Same as above, but also suggests what records will be taken.	
Theory, principles, and concepts	0	No conceptual side is identified.	
	1	A few concepts are identified, but without principles and theory OR a principle written is the knowledge claim sought in the laboratory exercise	
	2	Concepts and at least one type of principles (conceptual or methodological) or concepts and relevant theory are identified.	
	3	Concepts and two types of principles are identified, OR concepts, one type of principle, and a relevant theory are identified.	
	4	Concepts, two types of principles, and a relevant theory are identified.	
Records/ Transformations	0	No records or transformations are identified.	
	1	Records are identified, but are inconsistent with the focus question or the major event.	
	2	Records OR transformations are identified, but not both.	
	3	Records are identified for the major event; transformations are inconsistent with the intent of the focus question.	
	4	Records are identified for the major event; transformations are consistent with the focus question and the grade level and ability of the student.	
Knowledge claims	0	No knowledge claim is identified	
	1	Claim is unrelated to the left-hand side of the Vee.	
	2	Knowledge claim includes a concept used in an improper context OR a generalization that is inconsistent with the records and transformations.	
	3	Knowledge claim includes the concepts from the focus question and is derived from the records and transformations.	
	4	Same as above, but the knowledge claim leads to a new focus question.	
Total Score _____ / 18			

question; objects/events; theory; principles; and concepts, records/ transformation, and knowledge claims. Each topic is subdivided into four to five items. The topics, subtopics, and the points on each of these subdivisions are provided in Table 7.1. The purpose of using the rubric was to ascertain that students were proficient in constructing the Vee Heuristics during the first week of the study.

Test of Physics Achievement

The TPA was adopted from Glenbrook-K12 (2002). This achievement test was designed to identify students' conceptions about voltage, current, and resistors, the topics covered in the instruction unit involved in this study. The test is composed of five main parts as follows: (1) Part A includes 16 true/false type questions, (2) Part B includes 15 multiple choice questions, (3) Part C includes 5 diagrams for analysis (Students construct or analyze circuits), (4) Part D includes one question which requires students to relate variables, and (5) Part E includes two problems to be solved.

According to Novak (1981) "items at the application level of Bloom's taxonomy or above would measure meaningful learning" (p. 13). In other writings Ausubel, Novak, and Hanesian (1978) and Novak (1981) contend that a test must include items at the comprehension level and above to ascertain that a learner shows meaningful learning. As a result, the cognitive levels of problems or questions that measure student achievement after using the Vee Heuristic should all be at higher levels of blooms taxonomy, namely comprehension, application, analysis, synthesis, and evaluation. However, to ensure that an intervention does not influence students' achievement at lower levels negatively, the instrument should also include a number of items at the knowledge level.

To ensure that the TPA (Glenbrook-K12, 2002) was aligned with the curriculum, the test was analyzed by using a table of specifications. Additionally, a detailed description of the six levels of Bloom's taxonomy (Bloom, 1969) was used to analyze the cognitive levels of the test items. The analysis involved two experienced and certified physics teachers who were asked to classify the test items and match them with objectives. The analysis involved (1) the objectives used to design the lesson plans and test, and (2) a detailed description and examples of Bloom's taxonomy. Analyses of the test showed that 17% of the items were at the knowledge level, 14% at the comprehension level, 21% at the application level, 21% at the analysis level, 17% at the synthesis level, and 10% at the evaluation level. Finally, the analysis showed that the items were fully aligned with the objectives. The reliability (Alpha coefficient) of TPA was found to be 0.75. (Appendix A presents examples of the items used in TPA).

Physics Attitude Scale

Students' attitude toward physics was measured by using the PAS adapted from the modified Fennema-Sherman attitude scale. The original scale was developed by Fennema and Sherman in 1976 in order to measure students' attitudes towards mathematics. The scale used in this study consists of 47 items in the form of positive and negative statements. Participants respond to these statements on a 5-point Likert scale where A is "strongly agree," B "agree," C "not sure," D "disagree," and E "strongly disagree." The items are randomly arranged. The scale consists of four subscales: a confidence subscale (12 items), a usefulness subscale (12 items), a teacher perception subscale (12 items), and a subscale that measures physics as a male domain (11 items). The confidence subscale measures the confidence to learn and to perform well on physics tasks. The usefulness subscale measures beliefs about the usefulness of physics and its relationship to students' future education. The teachers' perceptions subscale measures students' perceptions of their teachers' attitudes toward them as learners while the male subscale scale asks about gender differences towards science. The highest possible scores for the confidence, usefulness, and teacher's attitudes subscales is 60 while that of the male domain subscale is 55. The reliability (Alpha coefficient) of PAS was 0.84 when given prior to the intervention in comparison to 0.92 following the intervention (Appendices B and C present examples of the items of PAS).

Physics Laboratory Attitude Questionnaire

Students' attitude towards laboratory work was measured by using an instrument developed by Yesilyurt (2004) to identify students' attitudes toward the physics laboratory. This questionnaire consists of 27 items and uses a 5-point Likert scale where 5 is "agree," 4 is "sometimes," 3 is "undecided," 2 is "partly agree," and 1 "never." The reliability (Alpha coefficient) of the questionnaire was 0.84 when given prior to the intervention in comparison to 0.89 when given following the intervention (Appendix C presents examples of the items of the Physics Laboratory Attitude Questionnaire).

Individual Interviews

Individual semistructured interviews were conducted with the experimental group students who answered 10 open-ended questions after the completion of the study. These questions were designed to investigate students' attitudes towards using the Vee Heuristic in laboratory sessions (Table 7.3).

**Table 7.3.   The Questions Used to
Interview Students in the Experimental Group**

1.	What was the part of the Vee Heuristic that you found helpful?
2.	What is the part of the Vee Heuristic that you found not helpful?
3.	Did the Vee Heuristic help you to understand the work in the lab?
4.	Did you notice the link between classroom explanation and the lab work through the Vee Heuristic? Describe these links.
5.	Would you like to use the Vee once again in the physics lab?
6.	Did you think that the Vee Heuristic helped you to perform better in physics why or why not?
7.	Did you enjoy using the Vee Heuristic?
8.	Is it hard to use the Vee Heuristic?
9.	Would you consider using the Vee Heuristic in studying for courses other than physics?
10.	Based on your experience what changes would you make to improve the use of Vee Heuristic?

Data Analysis

Data was analyzed based on the three research questions. In order to answer the first research question: Does using the Vee Heuristic in physics laboratories, in contrast to using verification-type laboratories, enhance students' meaningful learning of the target physics concepts? Student achievement was measured by using the TPA. A t test and a Multivariate Analysis of Variance (MANOVA) were used to test the existence of differences between the control and experimental groups. Moreover, a two-way analysis of variance (two-way ANOVA) was used to investigate the interaction between students' achievement levels and overall achievement on the TPA for the experimental and control groups. Three other two-way ANOVAs were used to investigate the interaction between achievement levels and scores on the different levels of Bloom's taxonomy.

Regarding the second research question: Does using the Vee Heuristic in guiding laboratory instruction enhance students' attitude toward science and the science laboratory? Students' responses on the PAS and the Physics Laboratory Attitude Questionnaire were analyzed by using an Analysis of Covariance (ANCOVA) with the pretest scores as covariates. In addition, interview transcripts were analyzed to identify participants' perceptions regarding the use and usefulness of the Vee Heuristic. Finally, to answer the third research question: Are there gender differences in attitude and achievement between students using the Vee Heuristic in physics laboratories and those who use verification-type laboratories? Data from the TPA were analyzed by using a two-way ANOVA; In addition, a two-way ANOVA was performed to check whether there were gender main effects or group-gender interactions for the questions at different cognitive levels. Moreover, the data from the PAS and the Physics Laboratory Attitude Questionnaire were analyzed by using a two-way ANCOVA.

RESULTS

Before answering the research questions, it was necessary to ensure the ability of the experimental group students to construct Vee Heuristics efficiently. This issue was addressed in two ways. One was completed by scoring the Vee Heuristics constructed after each laboratory. The second was completed by a test which assessed students' knowledge of the components of the Vee Heuristic and the relationships between these components. The mean of the scores of the Vee Heuristic was 41.37, while the mean on the Vee Heuristic test was 38.06 out of 54. Results are shown in Table 7.4. The results indicated that all students, in general, mastered the Vee Heuristic since the mean scores on both measures were above 70%.

As shown in Table 7.4, the mean score of the female students in the experimental group on the Vee Heuristic was 43.09, while that of the male students was 39.92 out of a maximum score of 54. A t test for independent samples was carried out to test whether the male and female students in the experimental group differed significantly on the scores on the Vee Heuristic. Nonsignificant differences were found with $t = 2.05$ ($p > 0.05$). In addition, the mean score of the female students in the experimental group on the Vee Heuristic test was 39.27 while that of the males was 37.04 out of a maximum score of 54. A t test for independent samples was carried out to test whether the male and female students in the experimental group differed significantly on the scores of the Vee Heuristic test. Nonsignificant differences were found with $t = 0.687$ ($p > 0.05$). Since there were no significant differences, it can be assumed that the females and males in the experimental group are equivalent in their proficiency in using the Vee Heuristic.

Table 7.4.   Mean Scores and Standard Deviations of Males and Females on the Vee Heuristic

Sex	Scores on the Vee Heuristic			Vee Heuristic Test Scores		
	n	M	SD	n	M	SD
Male	13	39.92	2.13	13	37.04	9.39
Female	11	43.09	5.06	11	39.27	5.72
Total	24	41.37	4.02	24	38.06	7.84

Research Question One

The first research question aimed at examining the effect of using the Vee Heuristic on students' meaningful learning. Student achievement was measured by results on the TPA. The TPA included questions at all levels of Bloom's taxonomy, consequently, four scores were generated from the test. An overall score (total) as well as a score on the knowledge level items (BloomK), a score on the comprehension items (BloomC), and a third score on the application and above items (BloomASE). The means and standard deviations of students' scores on the test of physics achievement categorized by bloom's cognitive levels were calculated for later analysis. The scores include those on the knowledge (BloomK) with a maximum score of 7, comprehension (BloomC) with a maximum score of 6, and application and above (BloomASE) level questions with a maximum score of 39 along with the total score on the TPA (Total) with a maximum score of 100.

To determine if there were initial differences between the two groups, students' overall achievement in Grade 9 was acquired. The mean score for the experimental group was found to be 72.00, while that of the control group was found to be 70.00 out of a maximum score of 100. At t test for independent samples was used to find if there were differences between the two groups. Results of the t-test showed that there were no significant differences ($t = 0.74$, $p > 0.05$). Consequently, a MANOVA was used to test the existence of differences between the control and experimental groups. These results show that there were no significant differences between the control and experimental groups on any of the variables. Assumptions for conducting MANOVA (homogeneity of variance and covariance and normality of distributions) were checked and found to be satisfied.

Students in all grades are divided by the school into different achievement levels. According to the school, students are categorized as follows: 91-100, Excellent; 81-90, Very good; 71-80, Good; 61-70, Fair; and 40-60, Poor. The number of students in each of those categories was identified based on the scores of Grade 9 and the means on TPA for each achievement level were calculated. However, because only a few students were categorized as excellent, they were included in the "very good" group (Table 7.5). A two-way ANOVA was used to investigate the interaction between students' achievement levels and overall achievement on TPA for the experimental and control groups. No significant interactions were found. Moreover, 2 two-way ANOVAs were used to investigate the interaction between achievement levels and scores on the different levels of Bloom's taxonomy (BloomK, BloomC, and BloomASE). No significant differences were found.

Table 7.5. Means and Standard Deviations of Students' Scores on the TPA by Achievement Level

	Control			Experimental		
Achievement levels	N	M	SD	N	M	SD
Excellent/Very good	4	68.84	15.6	5	69.23	6.61
Good	7	62.96	17.55	8	66.34	7.86
Fair	9	55.12	6.68	9	59.40	6.60
Poor	3	50.25	8.91	2	50.00	1.08

Research Question Two

To answer the second research question data from students' responses on the PAS adapted from the Modified Fennema-Sherman Attitude Scale, the Physics Laboratory Attitude Questionnaire developed by Yesilyurt (2004), and interviews with the experimental group students were used.

The Physics Attitude Scale

The PAS was given to students as a pretest and a posttest. The mean score for the experimental group on the pretest was 181.33, while that of the control group 175.82 out of a maximum possible score of 235. The mean score of the experimental group on the post-test was 182.21, while that of the control group was 157.09 out of a maximum possible score of 235. Table 7.6 presents the means and standard deviations of the control and the experimental groups.

ANCOVA, with the scores of post-PAS as the dependent variable and the pretest scores as the covariate, was performed to investigate the effects of using the Vee Heuristic in the physics laboratory on students' attitudes toward physics. Assumptions for conducting ANCOVA, linearity of the relationship between the covariate and dependent variable and homogeneity of regression, were checked and found to be satisfied. Results

Table 7.6. Means and Standard Deviations of Students on the pre- and Post-PAS

	Control			Experimental		
Questionnaire	n	M	SD	N	M	SD
Pre	22	175.82	23.33	24	181.33	17.20
Post	22	157.09	25.40	24	182.21	19.10

Note: One of the control students did not complete the questionnaire

showed that there were significant differences between the experimental and control groups ($F = 13.76$, $df = 1$, $p < 0.001$). Inspection of the means presented in Table 7.6 showed that the mean score of control group students dropped significantly while that of the experimental group did not seem to change.

The Physics Laboratory Attitude Questionnaire

The Physics Laboratory Attitude Questionnaire was given to students as a pretest and a posttest to assess the effect of participation in the Vee Heuristic laboratories on students' attitude toward physics laboratories. The mean score of the experimental group on the pretest was 105.75, while that of the control group was 104.04 out of a maximum possible score of 135. The mean score of the experimental group on the posttest was 105.17, while that of the control group was 91.52 out of a maximum possible score of 135. Table 7.7 presents the means and standard deviations of the control and the experimental groups.

An Analysis of Covariance (ANCOVA) with the scores of post Physics Laboratory Attitude Questionnaire as the dependent variable and the pretest scores as the covariate to investigate the effects of using the Vee Heuristic in the physics laboratory on students' attitudes toward physics laboratories showed that there were significant differences between the experimental and control groups ($F = 4.40$, $df = 1$, $p < 0.05$). Inspection of the means presented in Table 7.7 shows that the mean score of control group students dropped significantly while that of the experimental group did not seem to change.

Individual Interviews

Students in the experimental group answered 10 questions in a semistructured interview conducted after the completion of the study. These questions were designed to collect qualitative data about students' attitudes toward using the Vee Heuristic in the physics laboratory.

Table 7.7.   Means and Standard Deviations of the TPA by Gender

			Group				
			Control			*Experimental*	
		n	*M*	*SD*	*N*	*M*	*SD*
Gender	Male	14	60.60	12.04	13	64.55	6.84
	Female	9	57.17	15.17	11	61.11	6.84

Results of analysis of the students' responses to the questions are reported below.

When the students were asked about the component of the Vee Heuristic that they found helpful to learn physics in general, 14% said that the Vee Heuristic as a whole was helpful, while 23% and 19% consecutively of students found the "Records" and "Transformation" as the most helpful parts. The rest, 34%, had no opinion about the components that were helpful. However, when the students were asked about the components that were not helpful, 21% claimed that the "Objects" component was not helpful, while 31% said that all parts were helpful. The remaining 48% had no opinion about the components that were not helpful. Finally, 70% of the students said that they liked the Vee Heuristic in general, 5% did not like it at all, while the rest, 25% said that they felt neutral toward it.

All the students except one said that the Vee Heuristic was helpful for them to understand the work required in the physics laboratory and suggested that they would use the Vee Heuristic in their future work in physics laboratories. Eighty percent of the students claimed there was a clear link between the conceptual and the experimental sides of the Vee Heuristic. Moreover, they recognized a link between the conceptual side and of the Vee Heuristic and work in class and another link between Laboratory work and the experimental side of the Vee Heuristic. However, 20% of the students did not see the relationship between classroom explanations and the use of the Vee Heuristic in the laboratory. The following excerpt illustrates what one student said: "The Vee makes the student to work and to understand more the subject he is taking since he links between working (experiments) and theory (classroom)" (Student M. Z.).

Ninety percent of the students said that the Vee Heuristic was not hard to use but that it needed time at the beginning. One student claimed that "it was hard at the beginning since it was a new technique we are using and since we have to do all the parts alone without any help from the teacher" (Student R. A.). The rest, 10% of the students, found the Vee Heuristic difficult to use.

Eighty percent of the students replied that they would like to use the Vee Heuristic in other science subjects, while 20% said that they would not. When those who said that they would not use it were asked about the reason they replied that in their opinion physics is more open to laboratory work than chemistry and biology. One student said "I will use it if the biology or the chemistry teacher allowed me to do so" (Student N. S.).

To summarize, many students liked the Vee Heuristic and said that they thought they would learn more in physics when using this technique.

Moreover, most students saw a link between the different components of the Vee Heuristic and with class and laboratory work. Finally, most students found the Vee Heuristic easy to use but said that they were not ready to use it in subjects other than physics.

Research Question Three

The third research question of the study involved gender differences in attitude and achievement between students using the Vee Heuristic in physics laboratories and those who use traditional verification-type laboratories. To answer this question data from the TPA, the PAS, and the Physics Laboratory Attitude Questionnaire were used.

Gender Effect on Physics Achievement

To investigate group-gender interactions on students' performance on the TPA, a two-way ANOVA was conducted with gender and group as independent variables. Assumptions for conducting the ANOVA (equality of variance and normality of distributions) were checked and found to be satisfied. The means of males and females on the TPA are presented in Table 7.8. Table 7.8 shows that the mean of the females in the experimental group was 61.11 and 57.17 in the control group, while the mean of the males in the experimental group was 64.55 in comparison to 60.60 in the control group out of a maximum possible score of 100. Results show that there was no significant main effect or gender x group interaction. Results are shown in Table 7.9.

In addition, a two way ANOVA was performed to check whether there were gender main effects or group-gender interactions for the questions at different cognitive levels. For the knowledge level questions, the mean score of the male experimental group students was 5.84 while that of the females was 5.81 out of possible maximum score of 7.

Table 7.8.   Two-way Analysis of Variance for Group and Gender on the PAT

Source	SS	df	MS	F
Group	134.40	1	134.40	1.04
Gender	177.72	1	177.72	1.38
Group X Gender	4.40E-04	1	4.40E-04	0.00
Error	5518.34	43	128.33	

Table 7.9.   Means and Standard Deviations of Students on the pre- and Post-Physics Laboratory Attitude Questionnaire

		Control			Experimental	
Test	n	M	SD	N	M	SD
Pre	23	104.04	15.59	24	105.75	13.70
Post	23	91.52	22.13	24	105.17	22.13

Table 7.10.   Means and Standard Deviations of Students Scores on the TPA for Bloom's Knowledge Level (BloomK) by Gender

		Group					
		Control			Experimental		
		n	M	SD	N	M	SD
Sex	Male	14	5.35	1.00	13	5.84	0.68
	Female	9	5.55	1.13	11	5.81	0.87

Table 7.11.   Two-way Analysis of Variance for Group and Gender on the TPA for the Knowledge Level Questions (BloomK)

Source	SS	Df	MS	F
Group	8.29E-02	1	8.29E-02	0.09
Gender	1.61	1	1.61	1.88
Group X Gender	0.14	1	0.14	0.17
Error	36.76	43	0.85	

The mean score of the male control group students was 5.35, while that of the female students was 5.55. Results are shown in Table 7.10 above. Results show that there was no significant main effect or gender x group interaction (Table 7.11 above).

The mean score of the male students in the experimental group on the comprehension level questions (BloomC) was 3.92, while that of the females was 3.54. The mean score of the male students in the control group was 3.35, while that of the females was 3.66 out of a maximum possible score of 6. Results are shown in Table 7.12. No significant main effect or gender x group interaction was found at the comprehension level (Table 7.13).

Table 7.12.   Means and Standard Deviations of Students Scores on the TPA Divided According to Bloom's Comprehension Level

		Group					
		Control			Experimental		
		n	M	SD	N	M	SD
Sex	Male	14	3.35	1.49	13	3.92	1.55
	Female	9	3.66	1.32	11	3.54	1.21

Table 7.13.   Two-way Analysis of Variance for Group and Gender on the TPA at the Comprehension Level Questions.

Source	SS	Df	MS	F
Group	1.32E-02	1	1.32E-02	0.007
Gender	0.56	1	0.56	0.27
Group X Gender	1.34	1	1.34	0.66
Error	86.86	43	2.02	

Table 7.14.   Means and Standard Deviations of Students Scores on the TPA for Bloom's Application and Above Level by Gender

		Gender					
		Control			Experimental		
		n	M	SD	N	M	SD
Gender	Male	14	21.96	5.55	13	22.40	4.26
	Female	9	18.72	7.71	11	21.00	2.36

Table 7.15.   Two-way Analysis of Variance for Group and Gender on the TPA for Bloom's Application and above Level Questions by Gender

Source	SS	Df	MS	F
Group	62.11	1	62.11	2.31
Gender	21.37	1	21.37	0.79
Group X Gender	9.44	1	9.44	0.35
Error	1151.71	43	26.78	

Table 7.12 shows that the mean score of the male experimental group students on the application level and above questions (BloomASE) was 22.40, while that of the females was 21.00. The mean score of the male control group students was 21.96, while that of the females was 18.72. There was no significant main effect or gender x group interaction at the comprehension level (see Table 7.15).

Gender Effects on Attitude Toward Physics

To investigate group by gender interactions on students' performance on the PAS, a two-way ANCOVA was conducted with gender and group as the two independent variables and pretest scores on the PAS as covariate. The means of males and females on the PAS for the control and the experimental groups are presented in Table 7.16. The mean of the females in the experimental group was 177.18 and 170.33 in the control group, while the mean of the males in the experimental group was 186.46 in comparison to 147.92 for the males in the control group, out of a maximum possible score of 235. Table 7.17 shows that there was a significant group main effect and gender x group interaction ($F = 6.04$, $df = 1$, $p < 0.05$) with the attitudes of males in the experimental group improving more than those of females.

Gender Effects on Attitude Toward Physics Laboratory Attitude Questionnaire

To investigate group by gender interactions on students' performance on the Physics Laboratory Attitude Questionnaire, a two-way ANCOVA was conducted with gender and group as the independent variables and performance on the pre-Physics Laboratory Attitude Questionnaire as covariate. The means of males and females on the Physics Laboratory Attitude Questionnaire for the control and the experimental group are presented in Table 7.18. Female means in the experimental group were 101.00 and 88.50 in the control group, while the mean of the males in the experimental group was 108.69 in comparison to 93.85 in the control

Table 7.16. Mean Scores and Standard Deviations of Males and Females on the Post-PAS

		Group					
		Control			*Experimental*		
		n	*M*	*SD*	*N*	*M*	*SD*
Sex	Male	13	147.92	10.64	13	186.46	17.59
	Female	9	170.33	34.57	11	177.18	20.50

Table 7.17.   Two-way ANCOVA for Group and Gender on the Post-PAS

Source	SS	Df	MS	F
Group	4678.99	1	4678.99	12.53*
Gender	237.11	1	237.11	0.63
Group X Gender	2255.20	1	2255.20	6.04*
Error	15300.80	41	373.190	

*$p < .05$.

Table 7.18.   Mean Scores and Standard Deviations of Students on the Post-Physics Laboratory Attitude Questionnaire

		Group					
		Control			Experimental		
		n	M	SD	N	M	SD
Gender	Male	13	93.85	14.12	13	108.69	21.42
	Female	10	88.50	30.22	11	101.00	21.01

Table 7.19.   Two-way ANCOVA for Group and Gender on the Post-Physics Laboratory Attitude Questionnaire

Source	SS	Df	MS	F
Group	2062.82	1	2062.82	4.33*
Gender	412.70	1	412.70	0.86
Group X Gender	0.35	1	0.35	0.001
Error	19970.21	42	475.48	

*$p < .05$

group out of a maximum possible score of 135. Analysis showed that there was a group main effect ($F = 4.33$, $df = 1$, $p < 0.05$) and no group x gender interaction. Results are shown in Table 7.19 above.

DISCUSSION

Vee Heuristic and Achievement

Using the Vee Heuristic in the Grade 10 physics laboratory was expected to result in improved overall achievement in physics as well as achievement on higher level cognitive questions. These expectations were based on the assumption that the Vee Heuristic helps students to appreciate the interplay between what they already know and the new knowledge they are attempting to understand. Such interplay provides students with a framework to help them construct their knowledge about physics concepts and the opportunity to understand how scientific knowledge is developed through the process of reflecting on what they know and the investigations they undertake. This occurs as students reflect on the conceptual and methodological sides of the Vee Heuristic as they try to investigate their focus question.

However, results showed that there were no significant differences between the two groups even though the mean score on the TPA of the experimental group was slightly higher than that of the control group. The same pattern could be identified for students' mean scores on questions at the knowledge, comprehension, and application-and-above levels, where the differences were not significant but the experimental group scored higher than the control group. Moreover, students' results in the different achievement groups were almost similar for the experimental and control groups.

Results of the present study are not consistent with those obtained in the few previous studies that examined the effect of Vee Heuristic on students' achievement. For example, using the Vee Heuristic was found to result in higher achievement in science in the studies conducted by Soyibo (1991), Esiobuo and Soyibo (1995), and Lehman, Carter, and Kahle (1985). This lack of consistency between the present study and the other studies could be attributed to five reasons. First, the three studies presented above investigated the effect of Vee Heuristic on achievement in biology rather than physics. Moreover, these studies use both the Vee Heuristic and Concept maps and consequently, it is not possible to isolate the effect of using the Vee Heuristic by itself. A second possible reason for the difference in results could be attributed to the different conceptual structures of biology and physics taught at the intermediate school level, an area that is open for future research on the use of the Vee Heuristic in specific content areas. Third, it is possible that the achievement test did not address competencies empathized in the Vee Heuristic laboratory but rather on general competencies that could be acquired in any type of laboratory. Fourth, the lack of significant differences could be attributed to the small sample size. This result might point to the possibility of differences that were not picked up by the statistical tests due to the small sample size. Finally, there is a possibility that the approach used in introducing the Vee Heuristic did not allow the students to use it meaningfully to learn the physics concepts covered by the test because of the emphasis on relationships between evidence and theory rather than on content.

Other analysis investigated the interaction between sex and the effect of using Vee Heuristic. Females who participated in the Vee Heuristic laboratory achieved high scores on the TPA. The differences however, were not significant. Moreover, other analyses investigated the interaction between sex and the effect of using Vee Heuristic for the questions at different cognitive levels. Males and females of both groups performed similarly on the knowledge and comprehension level questions. Regarding the application-and-above level questions, the mean score of the males and females in the experimental group were both high. These differences did not achieve statistical significance. The literature review conducted for the purposes of this study did not identify research that investigated the effect of gender on the use of the Vee Heuristic even though there is theoretical support for investigating this effect of gender on science. Consequently, there is need for more research to understand such an effect.

Vee Heuristic and Attitudes Towards Physics and Physics Laboratory

Even though there is reason to believe that involving students in scientific inquiry experiences, similar to the Vee Heuristic, can potentially affect students' attitudes toward science positively, the literature review did not specifically identify research studies that investigated the effects of using the Vee Heuristic on students' attitudes toward science. Results showed that the mean score of the experimental group on the post physics attitude questionnaire was higher than that of the control group. This was not because the attitude of students in the experimental group improved but rather due to the fact that the mean score of the control scores of the control group dropped significantly. It seems that students' attitudes toward physics were affected by the Vee Heuristic. Experimental students' positive attitudes towards physics stayed the same while a sharp drop was clear for students who did not use the Vee Heuristic. The results of the present study are consistent with those obtained by previous studies examining the effect of inquiry laboratory activities on students' attitudes towards science. For example, Ajewole (1991) found that students using the discovery method had more positive attitudes than those using the expository method. Similarly, Mao and Chang (1998) showed that students in an inquiry group developed more positive attitudes toward earth science than those in the control group. Similar results were found by Gibson and Chase (2002). It is clear that involvement in the Vee Heuristic laboratory activities provided students with opportunities to be involved meaningfully in their own learning, an involvement that resulted in a sustained interest in physics and consequently a positive attitude toward the subject.

Using the Vee Heuristic in the Grade 10 physics laboratory was expected to result in positive attitudes towards physics laboratory, a prediction that was verified as evident in the results. The pattern of results was similar to those found in the attitude of students toward physics. That is, experimental students' positive attitudes toward physics laboratory stayed the same while a sharp drop was clear for students who did not use the Vee Heuristic

Results of the present study are consistent with those obtained in the very few previous studies that examined the effect of Vee Heuristic on students' attitudes towards laboratory work. For example, Tessier and Nakhleh (1993) conducted a study to investigate how students might react to general chemistry laboratories if they used the Vee Heuristic. Two major trends were found regarding the Vee Heuristic: Students felt increased confidence in the laboratory after using the Vee Heuristic and felt that the Vee Heuristic was a viable method for learning from laboratory. In addition, Lebowitz (1998) conducted a research study that aimed to evaluate students' perceptions of a laboratory guided by the Vee Heuristic compared to laboratories guided by the verification-type format. Results of this study yield that most students believed that they did more thinking and learned when using the Vee Heuristic and said that they would try to use the Vee Heuristic again in their laboratory work. Results of Tessier and Nakhleh (1993) and Lebowitz (1998) are also consistent with the results obtained from analyzing the interviews with students in experimental group of this study.

Further analyses investigated the interaction between gender and the use of the Vee Heuristic on students' attitudes toward physics. The mean score of the females in the experimental group was higher than that of the females in the control group. In the same sense the mean of the males in the experimental group was higher than those of the control group. As such, females and males who participated in the Vee Heuristic laboratory achieved higher scores on the post-attitude questionnaire than those who participated in the verification-type laboratory sessions. Males in the experimental group achieved the highest mean. This result is consistent with previous studies that showed that females had less positive attitudes towards science compared to males at the intermediate and secondary levels (Nokari, 1998; Weinburgh, 1995; Simpson & Oliver, 1985; Tamir, Arzi, & Zloto, 1974). However, it is worth noting that while male students had more positive attitudes toward physics than females, it is clear that females in the experimental group had more positive attitudes than those in the control group. Consequently, the Vee Heuristic had a differential effect on males and females.

Analyses of the interaction between gender and the use of the Vee Heuristic on students' attitudes toward physics laboratory revealed that the mean of the males in the experimental group was significantly higher than that of the males in the control group. The mean of the females, in the experimental group was also significantly higher than those of the control group Further analyses showed that there was a group significant main effect and no group by gender interaction. It is possible that the Vee Heuristic laboratory activities provided all students with opportunities to participate meaningfully in laboratory work, leading them to develop positive attitudes. It seems that the fact that females typically have less positive attitudes toward physics was overshadowed by their increasing attitude toward laboratories.

CONCLUSIONS AND IMPLICATIONS

This study provided some insights into the effect of using the Vee Heuristic in the Grade 10 physics laboratory on students' achievement in physics and attitudes toward physics and physics laboratory. The use of the Vee Heuristic in physics laboratories in contrast to using the verification-type laboratories did not affect students' achievement and as a result did not enhance students' meaningful learning of the target physics concepts. This might be attributed to a variety of factors including the nature of the subject matter taught, the instructional methods used in teaching the Vee Heuristic, the instructional procedures used in the Vee Heuristic laboratories, and/or the nature of the achievement test. However, the study provided strong evidence regarding the effect on attitudes. Specifically, students' attitudes toward physics and physics laboratory were consistently high before and after the laboratory while those attitudes became less positive for students in the control group. Finally, the study revealed that there were no gender differences regarding achievement and attitudes towards physics laboratory, however, males had more positive attitudes toward physics than females because of being involved in the Vee Heuristic laboratory.

The results of the study showed that the Vee Heuristic affected students' attitudes toward physics and physics laboratory positively. Consequently, the Vee Heuristic may serve as a useful tool to maintain student's positive attitudes toward physics and physics laboratories. This is further strengthened by the fact that students suggested in the interviews that the Vee Heuristic was helpful and that it could be used in subject areas other than physics. Further recommendations for teaching could be derived from the students' responses to the interview questions. According to students, the Vee Heuristic needs time to be mastered, consequently a longer training period might be need to achieve positive achievement results. Using the Vee Heuristic requires extensive professional development because the nature of this tool requires in-depth understanding of methodological and theoretical issues of which many teachers might not be aware.

Finally, the fact that this study is one of the few studies that investigated the effect of using the Vee Heuristic in a physics laboratory suggest that more research is needed to investigate the source of strong theoretical support associated with using this heuristic to improve student learning. Furthermore, more research is needed to investigate the effect of the Vee Heuristic on the achievement and attitudes of students of different age groups that may be related to developmental factors associated with using this heuristic.

APPENDIX A

Examples of Items of Test of Physics Achievement

1. A miniature light bulb with a specific resistance is connected to a 1.5-Volt battery to form a circuit. If it were connected to a 6-Volt battery instead, its resistance would increase by a factor of 4. (True/ False)

2. In which of the following situations will the light bulb light? List all that apply.

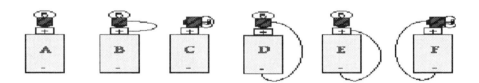

3. Birds can safely stand on high voltage electric power lines. This is because:

    (a) They are at low potential with respect to the ground.

    (b) They offer no resistance to current.

    (c) They always choose power lines that are not in use.

    (d) The potential difference between their feet is low.

    (e) They are perfect insulators.

    (f) They are perfect conductors.

4. The diagram at the right shows two identical resistors—R1 and R2 —placed in a circuit with a 12-Volt battery. Use this diagram to answer the next several questions.

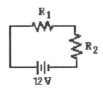

5. These two resistors are connected in.

    a. series   b. parallel   c. neither

6. If a third resistor (R3), identical to the other two, is added in series with the first two, then the overall resistance will ____ and the over-all current will ____.

    (a) increase, increase
    (b) increase, decrease
    (c) increase, remain the same
    (d) remain the same, increase
    (e) remain the same, remain the same
    (f) decrease, decrease
    (g) decrease, increase
    (h) decrease, remain the same
    (i) remain the same, decrease

7. Use proper schematic symbols to construct a diagram of a circuit powered by a 6-volt battery that consists of two 3-ohm resistors connected in series. Place ammeters in series at a location such that the current through each resistor can be measured and in a location such that the overall current in the circuit can be mea-sured. On the schematic diagram, use an unbroken arrow to indi-cate the direction of conventional current. Finally, indicate the ammeter readings on the diagram.

APPENDIX B

Examples of Items of PAS and
Physics Laboratory Attitude Questionnaire

Physics Attitude Scale (PAS)

A: strongly agree, B. agree, C. not sure about a question or you can't answer, D. disagree, and E. strongly disagree

1	I am sure that I can learn physics.	A	B	C	D	E
2	My teachers have been interested in my progress in physics.	A	B	C	D	E
3	Knowing physics will help me earn a living.	A	B	C	D	E
4	I don't think I could do advanced physics.	A	B	C	D	E
5	Physics will not be important to me in my life's work.	A	B	C	D	E

APPENDIX C

Physics Laboratory Attitude Questionnaire

5: fully agree, 4: agree, 3: undecided, 2: partially disagree, 1: fully disagree

1	I like working in the physics laboratory.	1	2	3	4	5
2	I would rather be told scientific facts than find them out from experiments at the lab.	1	2	3	4	5
3	I will use the knowledge learnt during physics laboratory in my life.	1	2	3	4	5
4	I feel comfortable when I solve new problems in physics laboratory.	1	2	3	4	5
5	Trying to understand physics laboratory experiments is a waste of time.	1	2	3	4	5

NOTE

1.   Corresponding author.

REFERENCES

Abd-El-Khalick, F., BouJaoude, S., Duschl, R., Lederman, N. G., Mamlok-Naaman, R, Hofstein, et al. (2004). Inquiry in science education: International perspectives. *Science Education, 88,* 397-419.

Ajewole, G. (1991).Effects of discovery and expository instructional methods on the attitude of students to biology. *Journal of Research in Science Teaching, 28*, 401-409.

Ausubel, D. (1963). *The psychology of meaningful verbal learning*. New York: Grunge & Stratton.

Ausubel, D., Novak, J., & Hanesian, H. (1978). *Educational psychology: A cognitive view* (2nd ed.). New York: Holt, Rinehart, and Winston.

Bloom, B. (1969). *Taxonomy of educational objectives: the classification education goals*. New York: McKay.

Brody, M. (1986).*Translating research reports into educational materials or how to take a neat piece of research and turn it into a curriculum*. East Lansing, MI: National Center for Teacher Learning. (ERIC Document Reproduction Service No. ED 273443)

Butler, B. (1999). Factors associated with students' intentions to engage in science learning activities. *Journal of Research in Science Teaching, 36*(4) 455-473.

Carillo, L., Lee, C., & Rickey, D. (2005).Enhancing science teaching by doing. *The Science Teacher, 72*(7), 60-64.

Esiobuo, O., & Soyibo, K. (1995). Effects of concept and Vee mapping under three learning modes on students' cognitive achievement in ecology and genetics. *Journal of Research in Science Teaching, 32*(9), 971-995.

Fennema, E., & Sherman, J. (1976). Fennema-Sherman Mathematics Attitude Scales. *Catalog of Selected Documents in Psychology, 6*, 31(Ms. No. 1225).

Freedman, P. M. (1997). Relationship among laboratory instruction, attitude toward science, and achievement in science knowledge. *Journal of Research in Science Teaching. 34*, 343-357.

Gardner, H. (1991). *The unschooled mind: How children think and how schools should teach*. New York: Basic Books.

German, J., Haskins, S., & Auls, S. (1996). Analysis of nine high school biology laboratory manuals: promoting scientific inquiry. *Journal of Research in Science Teaching, 33*(5), 475-499.

Gibson, L., & Chase, C. (2002).Longitudinal impact of an inquiry based science program on middle school students' attitudes toward science. *International Science Education, 86*, 693-705.

Glenbrook-K12. (2002). *Electric circuits review*. Retrieved November 20, 2007, from http://www2Glenbrook.K12.il.us/gbssci/phys/chemphys/reviews/u13review/u13ans1.html

Greenfield, T. (1997). Gender and grade level differences in science interest and participation. *Science Education, 81*, 256-276.

Gurley-Dilger, L. (1992). Gowin's Vee linking the lecture and the laboratory. *The Science Teacher, 59*, 50-57.

Haussler, P., & Hoffmann, L. (2002). An intervention study to enhance girl's interest, self-concept, and achievement in physics classes. *Journal of Research in Science Teaching, 39*(9), 870-888.

Heinzen-Fry, J., & Novak, J. (1990). Concept mapping brings long-term movement towards meaningful learning. *Science Education, 74*(4), 461-472.

Hofstein, A., & Lunetta, N. (1982). The role of the laboratory in science teaching: neglected aspects of research. *Review of Educational Research, 52*(2), 201-217.

Hofstein, A., & Lunetta, N. (2004). The laboratory in science education: foundations for the twenty-first century. *Science Education, 88*(1), 28-54.

Lebowitz, J. (1998). *Use of Vee maps in a college science laboratory.* East Lansing, MI: National Center for Research on Teacher Learning. (ERIC Document Reproduction Service No. ED 419694)

Lehman, J., Carter, C., & Kahle, J. (1985). Concept mapping, Vee mapping, and achievement: result of a field study with black high school students. *Journal of Research in Science Teaching, 22*(7), 663-673.

Lehrer, R., Schauble, L., & Petrosino, A. (2001). Reconsidering the role of experiment in science education. In K. Crowley, C. Schunn, & T. Okada (Eds.), *Designing for science: Implication from everyday, classroom, and professional settings* (pp. 251-277). Mahwah, NJ: Erlbaum.

Mao, L., & Chang, Y. (1998). Impacts of an inquiry teaching method on earth science students' learning outcomes and attitudes at the secondary school level. *ROCD, 8*(3), 93-101.

Mason, L. (1992). Concept mapping: a tool to develop reflective science education. *Science Education, 76*(1), 51-63.

Nakhleh, B. (1994). Chemical education research in the laboratory environment: How can research uncover what students are learning? *Journal of Chemical Education, 71*(3), 201-205.

Nakhleh, M., & Krajcik, J. (1993). A protocol analysis of the influence of technology on students' actions, verbal commentary, and thought processes during the performance of acid-base titrations. *Journal of Research in Science Teaching, 30*, 1149-1168.

Nataraj, M., & McManis, K. (2001, August). *Application of educational and engineering research to classroom teaching,* Paper presented at the International Conference on Engineering Education, Oslo, Norway.

Nokari, R. (1998, May). *An investigation of factors affecting students' attitudes toward science in Beirut area.* Unpublished master's thesis presented at the American University of Beirut, Lebanon.

Novak, J. (1981). Applying learning psychology and philosophy of science to Biology teaching, *American Biology Teacher, 43*(1), 12-20.

Novak, J. D. (2002). Meaningful learning: The essential factor for conceptual change in limited or inappropriate propositional hierarchies leading to empowerment of learners. *Science Education, 86*(4), 548-571.

Novak, J., & Gowin, B. (1984). *Learning how to Learn.* Cambridge, MA: Cambridge University Press.

Novak, J., Gowin, B. D., & Johanson, T. G. (1983). The use of concept mapping and knowing Vee mapping with junior high school science students. *Science Education, 67*(5), 625-645.

Okebukola, A., & Ogunniyi, B. (1984). Cooperative, competitive and individualistic science laboratory interaction patterns effects on students' achievement and acquisition of practical skills. *Journal of Research in Science Teaching, 21*(9), 875- 884.

Pharaon, A. (1984, July). *The development of an instrument that measures Lebanese fifth elementary students' attitudes toward science and the comparison of those attitudes by sex and language of instruction.* Unpublished master's thesis presented at the American University of Beirut, Lebanon.

Raad, H. (1997, February). *An investigation of gender differences in attitude toward science.* Unpublished master's thesis presented at the American University of Beirut, Lebanon.

Roehric, G., Lulie, A., & Edwards, M. (2001). Versatile Vee maps. *The Science Teacher, 68,* 28-31.

Roth, W. M. (1990). Map your way to a better lab. *The Science Teacher, 57,* 31-34.

Roth, W. M., & Bowen, M. (1993). The unfolding Vee. *Science Scope, 16,* 28-32.

Roth, W. M., & Roychoudhury, A. (1993). Using Vee and concept maps in collaborative settings. *School Science and Mathematics, 93*(5), 237-244.

Rubin, A., & Tamir, P. (1988). Meaningful learning in the school laboratory. *The American Biology Teacher, 50*(8), 477-482.

Salta, K., & Tzougraki, C. (2004). Attitudes toward chemistry among 11th grade students in high schools in Greece. *Science Education, 88*(4), 535-547.

Simpson, R., & Oliver, S. (1985). Attitude toward science and achievement motivation profiles of male and female science students in grades six through ten. *Science Education, 69,* 511-526.

Soyibo, K. (1991). Impacts of concept and Vee mapping and there modes of class interaction on students' performance in genetics. *Educational Research, 33*(2), 113-121.

Tamir, P. (1977). How are the laboratories used? *Journal of Research in Science Teaching, 14*(9), 311-316.

Tamir, P. (1989). Training teachers to teach effectively in the laboratory. *Science Education 73*(1), 59-69.

Tamir, P., Arzi, A., & Zloto, D. (1974). Attitudes of Israeli high school students towards physics. *Science education, 58*(1), 75-86.

Tessier, B., & Nakhleh, M. (1993, February). *Chemical education research in the laboratory environmen.* Paper presented at the annual meeting of the Hoosier Association of science teachers, Indianapolis, MN.

Trowbridge, L., & Bybee, R. (1990). *Becoming a secondary school science teacher* (5th ed.). New York: Merrill.

Von Secker, C., & Lissitz, R. (1999). Estimating the impact of instructional practices on student achievement in science. *Journal of Research in Science Teaching, 36,* 1110–1126.

Weaver, G. (1998). Strategies in K-12 science instruction to promote conceptual change. *Science Education, 82,* 445-472.

Weinburgh, M. (1995). Gender differences in student attitudes toward science: a Meta analysis of the literature from 1970-1991. *Journal of Research in Science Teaching, 32,* 387-398.

Weiss, I. (1987). *Report of the 1985-86 national survey of science and mathematics education.* Research Triangle Park, NC: Research Triangle Institute.

Yesilyurt, M. (2004). Student teachers' attitudes about basic physics laboratory. *The Turkish online Journal of Educational Technology, 3*(4), Article 7.

PART III

THE STATUS AND IMPACT OF TECHNOLOGY INTEGRATION IN LEARNING AND TEACHING SCIENCE

INTRODUCTION

The National Science Teachers Association (NSTA) placed great importance on the use of computers in the classroom. In its position statement, the NSTA noted:

> It is therefore the position of the National Science Teachers Association that computers should have a major role in the teaching and learning of science. Computers have become an essential classroom tool for the acquisition, analysis, presentation, and communication of data in ways, which allow students to become more active participants in research and learning. (NSTA, 2007, p. 1)

In fact, today's youth use the Internet for a "substantial stake in their social and educational lives" (North Central Regional Laboratory, 2007, para. 5).

While no one denies the need to use computers in the classroom, we must not forget that science educators, as well as science students, need additional knowledge and skills in the vast amount of other technology areas available for teaching and learning. The development of this knowledge and these skills encompass areas such as typical laboratory instrumentation (e.g., spectrophotometer), remote sensing, telecommunications, global positioning satellite technology, and data collecting probes and software.

Therefore, the third section of this volume offers the reader a view of current research into effective technology integration in science courses

The Impact of the Laboratory and Technology on Learning and Teaching Science K-16, pp. 203–205

and teacher education courses. In the first chapter of this section, Randall Davies, Constance Sprague, and Colleen New advocate the use of technology in problem based learning (PBL). In the next chapter, David Ucko and Kirsten Ellenbogen report on the status of research into informal science learning at site-based centers like museums and zoos as well as in media-based educational settings involving television, print, film, and the Internet. In the final chapter of this section, Craig A. Wilson describes the status of the research in the seamless integration of technology in elementary and secondary classrooms, as well as, teacher preparation programs.

In the opening section of chapter 8, Davies, Sprague, and New discuss the impact of the No Child Left Behind Act on assessment of science content and process skills, noting most standardized tests primarily assess science content knowledge. Also, the authors report that the legislation requires student use of technology in the learning process. To adequately assess science process skills, the authors advocate the use of PBL along with appropriate technologies used in inquiry lessons. Next, Davies, Sprague, and New outline the results of a case study in the use of technology (computers and electronic science probes) in conjunction with PBL in nine sixth grade classrooms. The findings indicated: (a) Teachers and students must have training in effective use of the technologies, (b) Teachers must have a good understanding of the nature of science to effectively use the technology in science teaching, (c) Student motivation was increased through the use of technology enhanced inquiry lessons, and (d) To effectively implement PBL and technology integration, teachers must have good classroom management and organizational skills.

In chapter 9, Ucko and Ellenbogen examine the role of technology in informal learning settings, museums and other out of classroom venues, through a review of the literature. After summarizing the literature on informal science learning in general, the authors address the impact of media, including film, television, and the Internet on science learning. At the conclusion of the chapter, Ucko and Ellenbogen discuss the current status of the research base on science learning in informal settings. While they note the difficulty in assessing learning in informal settings, they also advocate the need for additional quality research.

Wilson outlined the research in technology integration in preservice and in-service teacher education over the past 15 years in chapter 10. In the beginning of the chapter, Wilson reports that the impact of preservice and in-service teacher accreditation requirements, instituted in 1997 by National Council for the Accreditation of Teacher Education (NCATE), has resulted in a more seamless integration of technology in the classroom. This was contrasted with the typical utilization of technology, as an add-on to the curriculum, prior to 1997. Through a review of case studies,

Wilson describes a continuum of student-centered instruction with technology used in preservice education, from simply posting lecture notes on a Web site (limited student interaction) to a course requiring in-service teachers to prepare an Internet-based PBL project for their students to solve. Next, Wilson reviews case studies involving types of integration used in preservice education: science methods courses, science content courses, field-based experiences, and science teacher education programs. Finally, Wilson presents a summary of the findings of the review of literature and proposed the use of research-based solutions to increase technology integration in teacher preparation courses.

Part III begins by providing a summary of exemplary technology integration through the use of PBL in chapter 8 authored by Davies, Sprague, and New. In the next chapter of this section, Ucko and Ellenbogen reviewed the literature of technology integration research in informal learning, providing insight into the latest addition to informal learning, the Internet. In the final chapter of this section, Wilson provides insight into the current status of technology integration research in elementary and secondary classrooms and pre- and in-service teacher education programs.

REFERENCES

National Science Teachers Association. (2007). *NSTA position statement: The use of computers in science education*. Retrieved April 16, 2007, from http://www.nsta.org/positionstatement&psid=4

North Central Regional Laboratory. (2007). *Critical issue: Using technology to improve student achievement*. Retrieved April 16, 2007, from http://www.ncrel.org/sdrs/areas/issues/methods/technlgy/te800.htm

CHAPTER 8

INTEGRATING TECHNOLOGY INTO A SCIENCE CLASSROOM

An Evaluation of Inquiry-Based Technology Integration

Randall S. Davies, Constance R. Sprague, and Colleen M. New

Learning can be enhanced through the use of technology; however, learning does not take place simply because technology is used. Neither can technology fully replace the intelligent human guidance of a skilled science teacher. Technology improves science instruction in a laboratory situation when the learning activity is aligned with the learning objectives of the course, is well-structured, and when the equipment and materials to be used are well-managed. In order for this to happen the science teacher must have a clear understanding of the true nature of science, both the content and the process. Both teacher and students need to obtain a working knowledge of the function and utility of the technology. And while technology-enhanced instruction does tend to motivate students to participate in learning activities, learning is more likely to occur when the technology being used becomes a transparent tool in the learning process and not the main focus of the activity.

The Impact of the Laboratory and Technology on Learning and Teaching Science K-16, pp. 207–237

INTRODUCTION

The current Elementary and Secondary Education Act, commonly referred to as No Child Left Behind (NCLB), mandates an emphasis on technology integration in all areas of K-12 education from reading and mathematics, to science and special education (U.S. Department of Education, 2002). As a result, education leaders at the state and local levels are expected to develop plans that effectively employ educational technology into the curriculum. The question for most schools is not whether to integrate technology but how technology can best be employed to enhance student learning and increase student achievement as outlined by curriculum standards.

In science education, as it is for each subject, states are required to establish learning standards, or student academic achievement standards as they are referred to in NCLB (P.L. 107-110, Section 1111(b)(1)(A)); these standards are supposed to outline what students are to learn in each subject at each grade level; the standards are also to align with nationally recognized professional and technical standards (U.S. Department of Education, 2002). According to the *National Science Education Standards (NCES)* (National Research Council [NRC], 1996), science instruction should reflect the true nature of science, recognizing science as both a body of knowledge and a process. As a body of knowledge, science is the accumulation of all the concepts, facts, principles, laws, and theories we have come to know through investigation. As a process, science is a way of thinking about and investigating the world in which we live (Chiapetta & Koballa, 2002).

Largely due to the accountability demands of standardized testing, classroom-learning goals in science education have focused student learning primarily on the acquisition of content knowledge (Gross, 2005; Harmon, 1995; O'Neil, Sireci, & Huff, 2003; U.S. Department of Education, 2000). Only a small percentage of items on state standardized tests attempt to measure process skills, with these items typically measuring procedural knowledge rather than actual students' ability to conduct scientific inquiries (Harmon, 1995). Textbooks aligned to state standards are often used to help teachers accomplish science-learning objectives as they relate to the science content standards. Some teachers also utilize certain technologies in a variety of ways to help students learn the facts and theories, vocabulary and procedural knowledge that are typically tested on state standardized assessments. However, the development and practice of science process skills cannot be accomplished through textbook instruction alone. While some procedural knowledge is prerequisite, science process skills are best learned, practiced, and assessed in a laboratory setting (NRC, 1996; Rezba,

Sprague, McDonnough, & Matkins, 2007). In the science laboratory, using a scientific approach, students are expected to design and carry out investigations, utilizing appropriate technologies to accomplish their inquiries (NSTA, 2007).

Traditionally, schools have employed quite different pedagogical approaches when teaching science content knowledge and science process skills. Yet, for some, separating science content knowledge instruction from science process skill development is not a desirable instructional strategy; rather, these individuals believe that the processes of science should help to convey science content (National Science Teachers Association [NSTA], 2007). Teaching science content knowledge and developing students' science process skills can both be enhanced with the use of technology, but an understanding of the learning goals and the utility and function of the technology in accomplishing these goals is needed if the technology is to be used effectively.

REVIEW OF LITERATURE REGARDING FACTORS IMPORTANT TO EFFECTIVE SCIENCE INSTRUCTION

A review of current literature suggests that science instruction is enhanced through the use of technology. Technology alone, however, does not ensure quality instruction and improved student learning. Some key topics identified in the existing literature that impact science instruction include the importance of (1) having highly qualified teachers, (2) utilizing PBL strategies to enhance science instruction, and (3) the belief that technology integration will improve science instruction.

Highly Qualified Teachers and the True Nature of Science

A well-substantiated body of research links teacher quality and student learning (Darling-Hammond, 2000; Mendro, 1998; Stedman, 1997), with a particularly strong positive correlation between teacher subject area certification and student achievement (Goldhaber & Brewer, 1996). It should come as no surprise, then, that subject matter mastery is central to the definition of a "highly qualified teacher" under NCLB. As a part of the state certification process, highly qualified teachers must demonstrate competency in subject matter associated with the disciplines they teach. For teachers of science at any level, this includes knowing and understanding "the central ideas, tools of inquiry, applications, structure of science and of the science disciplines he or she teaches" (Interstate

New Teacher Assessment and Support Consortium [INTASC], 2002, p. 2).

A well-developed understanding of the structure of science, often referred to as "the nature of science," is critical content knowledge for teachers of science. The *NCES* include K-12 content standards calling for explicit teaching of the history and nature of science. Furthermore, all instruction has the potential to be informed—or misinformed—by the teacher's conception of the true nature of science.

Teachers of science must not only understand the nature of science but portray it in an authentic fashion to students so they become "scientifically literate." Simply put, scientifically literate students have knowledge of basic science concepts and processes; they are able to apply their knowledge and ability to better understand and explore the changing world in which they live (National Science Foundation, 1996).

So, what is science? Science can be thought of as having a dual nature; on one hand, science is a body of knowledge. It is what we know: the concepts, facts, principles, laws, and theories we have accumulated over time. We call this science "content" but must realize that it is dynamic in that it is always changing as new information and ideas come to light. Science is also a process. It is how we come to know what we know. This refers to the processes scientists and others employ as they attempt to investigate and learn new things. These processes include thinking in a way that relies on evidence for support and being skeptical of ideas that are not grounded in empirical evidence. The processes involve questioning and being inquisitive, looking for cause-effect relationships, and thinking things through using both inductive and deductive reasoning. In addition, the processes involve investigating with tools and materials and employing certain skills, often called science process skills, such as observing, predicting, inferring, measuring, hypothesizing, investigating, collecting data, and drawing reasoned conclusions (Rezba, Sprague, McDonnough, & Matkins, 2007).

Assessing Science Content Knowledge and Process Skill Development

Several long-term reform initiatives (i.e., the NSES Project, and the Scope, Sequence, and Coordination Project) emphasize the need for educators to create classroom-learning environments that encourage and promote students' literacy in science (NSTA, 2007). "Literacy," here, refers not only to students' understanding of the nature of science but also recognizing the relationship between science and society. Virtually all states have developed student academic achievement standards for science that include content knowledge and process skill development goals with the intent of having students become scientifically literate.

These standards serve as benchmarks or guidelines for what schools should teach and what will be assessed. However, as states attempt to develop assessments for accountability purposes these tests often do not assess content knowledge and process skills equally. Most state standardized science tests have a disproportionate number of content-based test items, with few items that assess growth or mastery of process skills (Harmon, 1995). Efforts to assess science process skills in any meaningful way on standardized tests are usually abandoned in favor of testing students' procedural knowledge of science investigation (Linn & Miller, 2005; Popham, 1999). This situation is understandable given the nature of learning outcomes being assessed. In other words, while a standardized test may be an effective and efficient way to measure science content knowledge, assessing science process skills must be measured through observation of individual performance, usually in an authentic laboratory setting (Linn & Miller, 2005).

Charged with meeting state standards and providing assessment data proving that standards have been met, schools must make curricular choices. Educators must choose the best instructional means and materials to maximize two very different but interdependent types of learning embedded in the science standards: content knowledge and process skills. For example, textbooks aligned to state standards, lecture based instruction, and providing access to science content information may provide the most efficient means for disseminating science content information to students and ensuring they will perform well on standardized tests. Conversely, process skills by nature are performance-based; they are best developed and assessed through well-designed inquiry-based investigations where students work in a laboratory setting. Content knowledge taught in the classroom utilizing a standards aligned textbook is often the method of choice because that is what is being tested, and that is what teachers and schools are being held accountable to teach.

There are those who suggest an alternative instructional approach. In the introduction to its position statement on the role of laboratories in science education the NSTA (2007) states:

> A hallmark of science is that it generates theories and laws that must be consistent with observations. Much of the evidence from these observations is collected during laboratory investigations. A school laboratory investigation (also referred to as a lab) is defined as an experience in the laboratory, classroom, or the field that provides students with opportunities to interact directly with natural phenomena or with data collected by others using tools, materials, data collection techniques, and models. (p. 3)

Throughout the process, students should have opportunities to design investigations, engage in scientific reasoning, manipulate equipment, record data, analyze results, and discuss their findings. These skills and knowledge, fostered by laboratory investigations, are an important part of inquiry—the process of asking questions and conducting experiments as a way to understand the natural world (NSTA, 2004). While reading about science, using computer simulations, and observing teacher demonstrations may be valuable, they are not a substitute for laboratory investigations by students (NRC 2006, p.3).

For science to be taught properly and effectively, labs must be an integral part of the science curriculum. The NSTA recommends that all pre-K-16 teachers of science provide instruction with a priority on making observations and gathering evidence, much of which students experience in the lab or the field, to help students develop a deep understanding of the science content, as well as an understanding of the nature of science, the attitudes of science, and the skills of scientific reasoning.

Problem-Based Learning

In an effort to integrate science content knowledge instruction and science process skill development, many secondary and middle school science programs have taken up an instructional approach called PBL as a means of meeting the *NSES* for multifaceted inquiry processes (NRC, 1996). In a PBL inquiry, students work collaboratively to identify a problem that is related to authentic local context or personal experience of interest to the students. The group then formulates questions about the problem that drives the investigation process toward a possible solution. The students attempt to answer the question together, learning the related content knowledge as they conduct scientific inquiries to gain a better understanding of the topic. Students use science process skills to question, predict, gather data, experiment, model, analyze, reflect, and present results. This approach is intended to simulate the conditions under which scientists work; it is intended to help students develop and practice science process skills while at the same time motivating them to learn and apply content knowledge. The teacher takes on a coaching role during PBL, allowing the students to be motivated by ownership of the problem, the direction of investigations, and the manner of documenting learning and solutions to problems (Torp & Sage, 2002). PBL is most effective when initial questions and solutions lead to additional questions and explorations, thereby reflecting the true nature of science. The PBL approach meets all of the requirements for inquiry-centered learning with some additional components highly valued by educational policymakers:

collaboration, learner-centered/learner-directed experimentation, and complex problems within an authentic context (NRC, 1996).

While there can be no perfect one-size-fits-all instructional model to develop all the dimensions of scientific literacy, the PBL model has great potential to target and integrate multiple standards for both content and process skill development. The pedagogical approach of PBL was developed over 30 years ago to simulate clinical problems and issues for medical students (Barrows, 1996; Barrows & Tamblyn, 1980). It is now being adapted for use in secondary and middle schools.

There are some caveats for educators who use PBL to develop scientific literacy in students. In any science exploration investigators attempt to systematically and rigorously explore messy, complex, and at times unwieldy problems. Often the answers are not easily attained and the ramifications to society are complex, with intelligent people often disagreeing on what a scientific result means and the best social policies to adopt. This certainly represents the conditions under which a scientist would work, but not everyone feels they are appropriate for the young science student (Lee & Songer, 2003). Without relevant foundational knowledge and experience, highly complex problems potentially create a disconnect between learner experience and the very knowledge the science instructor wishes to reinforce. Misconceptions about science concepts that come directly from learner experience are difficult to correct. A mismatch between the learners' abilities and the level of problem complexity can undermine science achievement (Chung & Chow, 2004; Crockett, 2004). Therefore, it is crucial that the teacher-as-coach guide learning, clarifying misconceptions and erroneous conclusions when they occur.

Still, the PBL method as an inquiry-centered learning approach is highly valued by educational policymakers in science education (NRC, 1996). This instructional approach has the potential to help students master both the content knowledge expected of them as outlined in the state standards and to help students develop important process skills needed to build new knowledge and better understand the world in which they live.

Technology Integration and Science Instruction

The integration of technology into the school curriculum is mandated by NCLB because it is commonly believed that learning is enhanced through the use of technology (U.S. Department of Education, 2002). However, not all share a common understanding of what technology is. For many, technology is synonymous with computer equipment, software, and other electronic devices; technology integration means using this equipment in the classroom. However, this is a very narrow definition of technology. *Educational technology includes any tool, piece of equipment or device, electronic or mechanical, which can be used to help students accomplish specified learning goals.* In science, tools often serve to enable or enhance observations and investigative processes. Useful technology may be simple, such as a magnifying lens, or it may be complex, such as a digital measuring instrument. These technologies often serve to save time or improve accuracy during an investigation. Sometimes the technology enables a laboratory investigation to take place, without which the learning activity would be relegated to a traditional lecture based classroom situation where students are told about the experiment or read about the science concepts in textbooks. However, learning does not take place simply because technology is used (Kleiman, 2000). For example, providing students with calculators does not ensure students will learn mathematics. The use of this valuable technology often results in improved (i.e., more accurate) calculations, but it can also have the unintended consequence of producing students with diminished conceptual understanding of the mathematical processes they are expected to know (Hembree & Dessart, 1986; Shockley, McGurn, Gunning, Graveley, & Tillotson, 1989). In order for the intended learning to occur, the technology must be effectively and purposefully utilized in a way that facilitates the intended learning objectives (Fisher, Dwyet, & Yocam, 1996; Kozma & Croninger, 1992; Lemke, 2005).

Summary of Literature Review

In the science classroom, the concept of a highly qualified teacher is based on teachers having an understanding of the true nature of science. Typically this includes having a mastery of the subject matter—understanding that science is a body of knowledge. However, it also includes understanding science as an investigative process. Teachers are expected to model the inquiry process for their students and provide opportunities for their students to conduct experiments. Many teachers, however, tend to focus their instruction on teaching content rather than process, as that is largely what is tested on state standardized tests. It is also believed that integrating technology into the science curriculum will enhance instruction. Unfortunately, many teachers have a very narrow definition of technology, relying heavily on computer technology.

RESEARCH PURPOSE AND QUESTION

This study presents findings and observations from a 3-year evaluation of a project which integrated technology into sixth grade science classrooms utilizing a PBL approach. The project was federally funded by a Transforming Education Through Technology Ed Tech grant. This initiative was one way in which the school corporation was attempting to accomplish its technology integration goals while at the same time meeting state requirements for academic achievement in science. An important goal of the evaluation was to determine the utility and usefulness of technology provided for use within the PBL science classroom environment. The underlying research question answered by this evaluation was: What factors influence the effective integration of technology into inquiry-based science instruction?

Method

Sixth-grade students at six public middle schools and three nonpublic schools in the Indianapolis metropolitan area were involved as the experimental group in this project. Two large public schools that did not participate in the project were used as a comparison group. The initiative provided participating schools with technology, as well as teacher training and support intended to facilitate increased student achievement in scientific thinking, vocabulary, comprehension of scientific concepts, and listening and speaking skills.

Two specific technologies were introduced: laptop computers and electronic science probes. Each of the six participating public schools received a computer cart with a set of 15 laptop computers, wireless networking and access to the Internet, a printer, four sets of science probes, books, and supplies. The three nonpublic schools initially received two laptop computers, wireless networking and access to the Internet, a printer, two sets of science probes, books, and supplies. Additional technology was added each year, and some schools supplemented the available educational technology through other grants, their own resources, and with existing technologies in each school.

Professional development for this one-to-one science project consisted of regular training activities for participating teachers. The training provided through the project was designed to support implementation of the PBL model, align instruction with academic science standards, and help familiarize teachers with the utility of the specific technology provided through the grant. Professional development activities included summer training workshops each year, as well as fall and spring project seminars where participants worked with content experts and technology coaches in planning and implementing their PBL units using the technology provided.

Teachers and students received coaching support through regular contact from digital coaches. Master teachers (i.e., digital coaches) supported teachers as they embedded technology into their standards-based curriculum. Digital coaches attended professional development sessions alongside participating teachers, participated in distance learning and online discussions, helped teachers in their classrooms as they gathered and studied student data, planned and initiated PBL units, and facilitated student efforts to use technology to complete their PBL units. Technicians were also available as needed to provide technical support to teachers when their technology failed to function properly.

Findings for this evaluation were based on qualitative analysis of data collected through regular site visit observations in the classroom; participant teachers' responses to surveys; and interview information from teachers, project administrators, and consultants. Data collected from surveys were analyzed to determine systematic differences and trends between and within schools. Qualitative methods were used to analyze and interpret data from site visits and interviews. This included information regarding unintended outcomes. Analyses of these data were conducted to determine emergent themes and patterns.

Project Results and Analysis

Presentations of the evaluation results and reflective analysis used to answer the research question are provided in this section of the chapter. Data consists of rich descriptions of classroom observations and interview transcripts followed by an analysis of the various issues. Several identifiable factors important to the effective integration of technology in the science classroom are explored. This is provided to the reader as background information and foundation for the evaluation's conclusions.

Equipment Utility

During one classroom visit, while waiting in the science lab for students to arrive, it was noted that a considerable amount of science equipment was stored neatly around the room and in storage closets. This equipment included a wide variety of technology the school had accumulated over the years, not the technology provided through this project. Some of this equipment remained unopened in its initial packaging and little of the equipment was ever used during any of our visits. Some of the equipment (e.g., tuning forks) had a very specific function, while other equipment (e.g., glassware) had more generic utility. (site visit observation)

In a laboratory situation, the equipment one uses is determined by the goals of the task at hand and the usefulness of the available equipment in facilitating the inquiry. Clearly, not all educational technology is useful for every class activity. When asked about the equipment provided to them through this project, participating teachers typically indicated that both the science probes and the computer equipment were somewhat useful in their science instruction. However, teachers clearly felt the computer technology, including wireless network and Internet connection, was much more useful to their overall instruction in their PBL unit than the accompanying science equipment. Teachers generally agreed that the science equipment had limited effect on the success of the overall PBL initiative. While the equipment was extremely useful and even essential in conducting specific science experiments, use of the science probes and other lab equipment was not always appropriate for exploring the problem selected by students in their PBL unit. As a result, only a few teachers indicated using the probes on a regular basis. Specific activities needed to be planned and prepared in which the probes or other equipment would be useful. It was additionally noted that for the probes to be most useful teachers often needed to purchase additional materials to facilitate data analysis and subsequent student understanding. For example, while the pH and electric conductivity probes measure some qualities of the water samples students tested (i.e., acidity and purity), additional materials such as testing strips, and additional experiments were often needed to explore exactly what was in the water and how these substances might have affected the usability of the water. These additional steps were rarely taken.

Students often used the computer equipment to obtain information through Internet searches; to create documents, presentations, and pamphlets; and to edit videos. Occasionally the laptops were used with science probe attachments to assist students in gathering and organizing data from experiments they were conducting. Some of the students used the computer to create concept maps and complete assignments. Still, much of the functionality and usefulness of the computer equipment might best be categorized as secondary to the central purpose of science learning goals related to developing process skills. Certainly students were using the technology in ways that may have facilitated the acquisition of science content knowledge, yet in many classrooms students seemed to be spending a considerable amount of their time and effort preparing presentations.

It should be noted that the use of a PBL approach in the science classroom, by design, will typically incorporate learning goals and objectives encompassing many subject areas not just those related to the science curriculum. When the PBL approach is used in the science classroom it attempts to simulate how scientists might solve problems within an authentic context, which will include learning objectives from a variety of subject areas (Torp & Sage, 2002). In several classes, students tended to spend a considerable amount of class time completing activities not directly associated with what might be considered significant science learning objectives. While the PBL questions and associated inquiries tended to be quite beneficial to student learning, typically only a small amount of time was spent actually conducting specific inquiries, analyzing the data, or exploring relationships between the inquiry data and established scientific principles. Most of the students' efforts were focused on gathering extant information and preparing presentations. This does not imply students were not learning important skills. It simply indicates that a large proportion of PBL instructional time was devoted to using technology to accomplish educational goals that might be considered secondary or peripheral to the development of science process skills and understanding the science content identified in the state standards. Computer equipment can be a very useful tool in creating presentations and acquiring information from a large number of sources, which may explain why teachers indicated the laptops and related infrastructure were used most often in their classrooms. Still, it could be argued that the focus of the instructional activities was often peripheral to the intended science learning goals or perhaps constituted an overemphasis on a few aspects of the science standards (i.e., presenting results).

Technology Integration and Motivation

Clearly just having technology available to them was a motivational factor for many students. However, individual students seemed more motivated to participate and most interested when they were the ones operating the equipment. Often groups of students sat around the equipment watching another student work with the equipment. Frequent discussion took place among students regarding who got to operate the equipment. One teacher also noted that if the computer was involved, students were more willing to take on the task than they would be if the task was to be done with just paper and pencil. (site visit observation and teacher comments)

In many ways, students tended to be enthusiastic about using the technology in the science classroom. Students holding, operating, or manipulating the equipment were focused on the equipment. For some students the motivation to participate seemed to stem from the fact that they were the ones using technology. If these students were not personally using the equipment, or once the equipment was no longer needed, their participation often diminished considerably. Clearly, students are motivated by technology and want to use it; however, in many instances "playing" with the technology tended to be a primary motivational factor of engagement for some students. For example, students comparing the thermal retention properties of paper and Styrofoam cups were provided handheld digital microscopes to use in their inquiry. This equipment allows students to view material close up and to capture digital pictures of the magnified images. Much of the initial use of the digital microscopes was spent viewing hair, skin, and other materials unrelated to the task at hand. Once students were finished playing with the equipment, they often abandoned it and proceeded to another activity. Few students, if any, considered using the equipment to compare the structural makeup of the two types of cups; instead, they tended to focus exclusively on using temperature probes to determine which cup was the best insulator by measuring and tracking temperature changes in liquids contained in two different types of cups as outlined by the task instructions. Those not operating the temperature probes tended to be engage in a variety of off task behaviors unrelated to the experiment. Only with considerable insistence from the teacher did the students attempt to explain the thermal retention differences in the two cups.

The integration of technology into schools is mandated by the current Elementary and Secondary Education Act because student learning can be enhanced through the use of technology (U.S. Department of Education, 2002). However, learning does not take place simply because technology is used (Kleiman, 2000). Certainly students seem motivated to use technology, but in educational situations, technology use tends not to produce an increase in learning unless the participating student's main intention is to learn (Davies, 2006). In addition, the technology must be used effectively to facilitate the learning expected of the students (Fisher, Dwyet, & Yocam, 1996; Kozma & Croninger, 1992; Lemke, 2005). The various technologies used in classrooms do not inherently cause learning to occur; rather these technologies, if integrated properly into instruction, can be used as tools to facilitate learning (Davies, 2006; Fisher, Dwyet, & Yocam, 1996). Integrating technologies into school environments has the potential to enhance learning, but schools must go beyond merely providing technology if students are to derive academic benefits.

Technology Training for Teachers

> None of the teachers we visited viewed the presence of technology in their classrooms as intrusive. None could even imagine "going back" to what they did before they had the technology. Most teachers viewed technology as just one more tool in the classroom and used it when it facilitated the learning activities they had planned. Still, not all teachers were equally adept at utilizing the technology provided. Over time, the gap in technological proficiency and innovation tended to grow between participating teachers. Less adept teachers tended to use technology in limited ways and often failed to seek out greater understanding of how the equipment worked, or how it might facilitate learning. In general, teachers tended to use equipment with which they were most familiar. (site visit observations and survey results)

It is not unusual for teachers to feel somewhat unprepared to use technology in their classrooms (U.S. Department of Education, 2003). This is particularly unfortunate since teachers who are well trained have been found to be more confident and tend to use technology more often in their instruction (Bray, 2005). As a result, training and ongoing support are considered essential aspects of any successful technology integration endeavor. Training might include helping teachers become familiar with new equipment and the instructional applications of the technology. Ongoing support includes both technology support (i.e., fixing equipment problems) and ongoing training situated in authentic environments (e.g., the classroom), so the knowledge and skills gained become less abstract and more meaningful (Glazer, Hannafin, & Song, 2005; Weston, 2005).

This initiative included both technical support and professional development training for the technology provided and the PBL approach. Overall the teachers felt the training and support were adequate, yet it was apparent that not all participants were equally determined to benefit from the opportunities made available to them. When faced with personal deficiencies in technological proficiency or the inevitable occurrences of technical difficulty, some teachers proactively and persistently sought out assistance. They practiced using the equipment on their own and paced themselves through the planned lesson just as students would do later. They anticipated where problems and glitches might lie, made notes about safety warnings to issue, and structured their class time to best accommodate the activity considering the learning goals and the limitations of time, space, and the equipment. As teachers became more comfortable with the capabilities of the technology they tended to find creative ways in which equipment might best be used to promote the intended learning goals. In comparison, teachers less willing to build their skills and understanding were often reluctant to use the technology on a regular basis, often incorporating the equipment in only one way or for only one specific experiment. When technological difficulties with the equipment occurred some teachers simply abandoned its use. Teachers with limited technological ability were more likely to leave equipment problems unresolved for extended periods of time, often restricting the equipment's usefulness and utility.

Technology Training for Students

> When teachers were unfamiliar with the technology they often relied on students to figure out on their own how the equipment worked. During one class session, students were observed trying to create a brochure outlining the information they had been asked to gather for their PBL unit. Students were often observed asking other students how to use the computer or how to do something with the software. Students were also observed trying to accomplish tasks in ways that could be done more quickly and efficiently if they had a better knowledge of the equipment and software capabilities.
>
> In another class, students were observed trying to complete an experiment using a force sensor. This science probe was one that was rarely used by any of the participating groups. Neither the teacher nor the students knew or could figure out how to use the equipment; eventually they gave up. (site visit observations)

Teachers were not the only ones who needed training in the use of technology. It is a common fallacy to suppose that because students are growing up in a technological age they are somehow instinctively more able to use technology to learn what is expected of them in school. Students today are no more or less capable of learning to use available technologies than students have been in the past. In fact, today's students typically use technology for social pursuits (i.e., communication and entertainment) but not necessarily for academic learning (Peck, Cuban, & Kirkpatrick, 2003). Students enjoy using technology, and many of them have good information-gathering skills, but often they lack sophistication in understanding and evaluating the information they retrieve (Stucker, 2005). Technology proficiency in some areas (e.g., Internet, e-mail, online games, or iTunes) often results in a deceptive sense of self-sufficiency. Educators sometimes mistake this confidence, or lack of inhibition, for skillfulness in using instructional technology for academic purposes. Too often these students have little or no working knowledge of new technologies being provided to them to facilitate their academic learning.

In order for students to effectively use equipment in their inquiries they must become familiar with the operation and function of the technology. Teachers with a good understanding of the equipment's operation and function tended to be better (i.e., more efficient and more thorough) at helping students learn to use the equipment. A common practice employed by teachers was to allow student time to "play" with the technology in order to become acquainted with its function and utility. This practice was more prevalent with the less technologically proficient teachers. As teachers became more personally proficient with the technology, they tended to provide demonstrations prior to allowing students time to get familiar with the equipment. This tended to reduce the amount of time required for this activity and enabled students as a group to start on the primary learning activities more quickly. In terms of instructional effectiveness and efficiency, a guided practice approach seemed to be much better than the self-discovery method.

In general, students spent a considerable amount of instructional time learning how to use equipment; moreover, this did not always result in students using the equipment efficiently and effectively to accomplish the primary learning objectives. The technology was often most useful once students got past the novelty of the equipment and determined how the equipment could help them complete their inquiry. Once this happened, the technology became transparent (i.e., almost invisible to the learning process). Equipment with which students were familiar was more likely to be used as a learning tool. Once students started focusing on the task of completing an assignment, using the technology was simply a way to get the data they needed to answer their questions or complete the tasks assigned to them.

Technology Use and Overcoming Misunderstandings

During a science lab, a group of students was observed using two temperature probes in a single beaker of water to measure temperature over time. While conducting the inquiry students noticed a discrepancy in the measurements. One probe gave a slightly different reading than the other probe. Reflecting on the result, the students concluded that one of the probes must be faulty, denoting a misconception that the temperature of the water in the cup is necessarily constant throughout.

In another situation, while observing students doing experiments with the pH and electric conductivity probes, it was noted that several students concluded that only neutral pH water was good to drink and that water with "stuff" in it was not good; in other words, it must be absolutely pure. One student tested a sample of tap water and found it to be slightly acidic with some "stuff" in it. When asked, the student concluded that the water was "not good, you shouldn't drink it." A moment later during cleanup the student proceeded to dispose of the water sample by drinking it as per the instructions given by the teacher. (site visit observation)

At times, a student's lack of foundational knowledge and experience tends to produce misconceptions about some science concepts. In elementary school settings like these (i.e., small group collaborative learning situations), problems sometimes arise when the teachers do not guide the learning in ways that focus students on the learning objectives, or fail to correct misconceptions and erroneous conclusions as they occurred. Technology can be a useful tool, but it will not correct misunderstandings that result from faulty logic or misinterpretation of the data; this requires human intellect. It seems that elementary students may need more direction and guidance, especially during the analysis phase of an inquiry. Technology facilitates the process of analysis by quickly generating, collating, and displaying data. But technology is no substitute for intelligent human guidance of a teacher during analysis. It is important not to let students form incorrect conclusions as a result of their analyses, as these misconceptions may be extremely difficult to correct later (Lee & Songer, 2003).

Technology Use and Building Understanding

It was noted that the computer equipment and the science probes used together were extremely useful at assisting student in making graphs and charts from experiment data. The quality of these artifacts was considerably better than what students might be able to produce by hand; sometimes the graphs and charts were created automatically from real time data. However, some of the students seemed to think that producing the graph or chart was the end product of the learning activity. (site visit observations)

To build knowledge and understanding using a PBL model, students must be engaged in activities that promote reflective analysis of the question being posed (Harwood, 2004; Jones, Valdez, Nowakowski, & Rasmussen, 1994). While graph construction, for example, is an important skill, interpretation and analysis of the data might be considered a higher order learning objective. Without the use of technology, students tended to spend considerable amounts of time organizing and presenting the data. Using technology can facilitate the ability of students to organize and present the information they collect; but the ability to analyze data, identify relationships, interpret the results, and draw conclusions based on the results cannot be done by the technology. Reflecting on the results is required if students are to make sense of the observations—in other words, to learn (Harwood, 2004;

White & Frederiksen, 1998). When the instructional task is structured appropriately, technology can be used to quickly get students to the point where they can spend time on higher order learning goals. When the instructional task is well-structured, technology can automate some of the time-consuming tasks of data collection and display, allowing students to focus on a specific higher order learning objective.

The Importance of PBL Question Selection

> The selection of a PBL problem varied considerably between classrooms from year to year. During the first year of the project, it was noted that most problems selected by students for the PBL unit tended to be in the area of Earth Science. Possibly as a result, there is evidence that students from the participating schools did better on the Earth Science portion of a standardized science achievement test than a comparison group of students. During subsequent years of the project, the PBL problems selected often focused more on a social issue rather than a scientific topic. Only a few of the questions selected lent themselves to students conducting their own experiments. (evaluation findings)

In science, there are two basic types of questions an investigator, or a team of student investigators, might pursue. One type of question can be answered by reviewing available literature: the researchable question. Students answer researchable questions by gathering and compiling enough existing information from credible sources until they feel confident that the question has been answered. An example of this type of question is: "What are the effects of substance abuse on society?" Answering this question would most likely require students to search the existing body of scientific knowledge for extant research. Computer equipment, the Internet, and related software programs are the most useful technology for this type of inquiry. Students are primarily acquiring content knowledge by doing this activity.

Another type of question is one that leads to experimentation. This type of question is sometimes called a testable question. In order to answer a testable question, an experiment is designed and conducted to test possible answers. Measurable data is collected, organized, analyzed and interpreted, and reported as a logical conclusion is drawn from the data. An example of a testable question is: "What effect does a lit light bulb have on the temperature of a room?" Students can design an experiment, manipulating the conditions to determine possible answers. A variety of science equipment and technology might be used for these types of inquiries. Students are primarily developing process skills by doing this type of activity.

While both types of questions have value in that they lead to new learning, it is only when students get involved in seeking answers to testable questions that they begin to model and understand what scientists do when they experiment. One important aspect of science is that it has an empirical nature (i.e., it is based on observable evidence). Another important aspect of science is that it is creative. When students design their own experiments, they are creating something new. Our observations suggested that teachers who were skilled at helping to guide students' selection of a PBL question toward a testable question, one that could be answered through investigation, while still enabling students to retain ownership of the question, were more likely to develop students' ability to think scientifically. For example, the PBL question, "How can we conserve energy at home?" led students to design their own investigation comparing the amount of heat given off and energy used by various wattages of incandescent and fluorescent light bulbs. Another PBL question, however, "Why is our population becoming obese?" did not lend itself as well to the gathering of empirical data. Students primarily gathered information they found from other sources to answer this question. Only a few schools selected problems that lent themselves to students conducting investigations through scientific experimental methods, although several groups conducted smaller investigations based on testable questions related to the primary PBL question.

Most of the PBL questions selected by the various schools tended to be researchable questions rather than testable questions. Students were often able to find answers to the PBL question they selected by searching existing sources and seeking out experts' opinions. Several of the questions selected for study might best be described as social issues. This is not inappropriate as these types of questions can be linked to the standards and students were quite enthusiastic about these topics; however, these types of questions are often very complex and the answers often depend on social values, personal morals, economic, and political factors. Few of the student presentations we observed focused much on the science concepts related to the "social issue" PBL questions selected. Given that standardized achievement tests focus more on content knowledge, and considering the age of the students, this level of understanding and detail may be developmentally and pedagogically appropriate. The knowledge and skills students gain from testable PBL explorations may not facilitate increased science achievement as measured by standardized assessments (Falk & Drayton, 2004). Certainly, teachers must guide student selection of a PBL problem so the learning aligns with the science concepts student will be tested on; alternately they must supplement the instruction with lessons that ensure coverage of the learning objectives prescribed in the state science standards. In fact, teachers participating in this project often reverted to traditional classroom instruction centered on textbook information to accomplish learning goals associated primarily with building content knowledge that students would be expected to know on the standardized tests.

The role the teacher assumes in getting students to ask and answer testable questions is quite different from the role assumed by a teacher involving students in asking and answering research questions. In a PBL approach the teacher facilitates the exploration and tries not to be the source of answers. When a teacher involves students in posing and answering testable questions, the selection of technology to support the investigation is also student centered. While the teacher provides the resources needed to conduct the exploration, students make decisions about how to accomplish a task, what materials are needed, and how to gather and organize data. Likewise, the analysis of the data and conclusions students draw are guided by an experienced teacher through questions rather than directions: What do these results suggest? In what way does this result support your hypothesis? Is there an alternative explanation? What additional inquiries might help you better understand the issue? Is that consistent with what we know about this topic or what experts say? The teacher guides the activity but does not control it (Torp & Sage, 2002). Using this approach helps students develop the ability to answer their own questions through inquiry and research rather than rely on the teacher to tell them the answers.

Laboratory Facilities and Organization Analysis

In the first year of the project, students were observed completing an inquiry involving endothermic reactions. Endothermic reactions absorb heat energy resulting in a loss of heat in the surroundings materials that can be observed. The activity is a standard science experiment associated with the science equipment provided through the IPS initiative. The equipment included detailed computer assisted instructions and the temperature probes needed to complete the task but no additional materials (i.e., water, Alka-Seltzer, beakers, etc.). The relatively small classroom doubled as a science lab with the lab scheduled in a 45-minute class period. The instructor was only somewhat familiar with the science probes as she had only recently received the science equipment, laptops, and associated software. Students were assigned to groups each with a laptop computer and a set of science probes. The computer program guided students through the lab giving background information, asking students to form a hypothesis, providing an equipment list of the necessary materials, and asking students to answer questions about the results. The students were well behaved and seemed eager to complete the experiment. As students got started they tended to skip the background information and made hypotheses like, "I think the water will bubble." As students got to the equipment list they set out on a search through cupboard drawers for the equipment. After a few minutes searching through equipment drawers, a few of the groups were unable to find large enough beakers and had to settle for smaller containers unsuitable for the experiment. The teacher busied herself helping students gain access to the laptops as the passwords weren't working, answering students' questions, and helping students look for equipment and materials. The classroom had no sink, so students asked permission to go down the hall to the janitors' room to obtain water. Unfortunately the supply of Alka-Seltzer tablets was forgotten at home, so fizzy soda pop was substituted in place of the Alka-Seltzer tablets, which resulted in no endothermic reaction occurring. One group began recording temperature over time but noted no change. None of the groups completed the task. As the class period was coming to an end students were asked to clean up and put the equipment away. The teacher indicated they would try again another day. (site visit observation)

The third year of the project we had the opportunity to visit a school where the teacher had become especially adept at the PBL approach. She too was required to conduct science labs in a small classroom without a sink and no storage space for equipment. However, the desks in the classroom were organized in clusters, leaving room for students to congregate on a carpet in the front of the room. During our visits we observed students involved in an inquiry looking at temperature change in a container of water. The lab was scheduled for a 45-minute class period. The problem was posed as a challenge, "Who can get the greatest degree of temperature change to occur?" Students had been working on the problem the previous week and would be completing the inquiry the following week. During the previous week, groups of students had formulated a plan based on a hypothesis then attempted to carry out their plan. At the end of that class students had requested specific materials and equipment they would need for the class we observed. A table filled with requested materials was set up in the room, including pitchers of water. Due to safety concerns, no flames or electrical heating apparatus could be used. The teacher started the class by bringing students together for a prelab discussion. She focused them on the task. She pointed out the location of the materials they had requested; these included ice, string, and various sizes and types of containers. Students set out collecting the equipment they had requested and were quickly carrying out their plans, recording and charting the temperature with laptops and temperature probes. They were obviously well acquainted with the equipment. Students were observed discussing the merits of various suggestions and why they felt their plan might work. Each of the groups completed at least one attempt to track the temperature change as they implemented their plan. The teacher engaged students, asking them to remember to take notes, and rather than directly answering questions when students asked her what to do, asking what the students felt was best and why. One girl was overheard exclaiming "I love science!" Toward the end of the class, students were asked to come together and discuss issues, concerns, and what they would need to complete the task the following class period before cleaning up. The teacher indicated to us that after the task was completed the following week the class would have a discussion about the various approaches and how they related to the concept of heat transfer and energy dispersion. (site visit observations)

Characteristics of the physical facilities in each of the participating schools varied. Some schools had two separate rooms which teachers could use as needed: one for the classroom and one for the science lab. A few schools had only a moderately small classroom that doubled as a science lab. Typically these rooms had little storage facilities for science equipment. One school held their classes in a large room with classroom and science lab configurations set up comfortably in the same room. This room also contained ample storage facilities with convenient access to science equipment and classroom materials. The physical facilities, however, did not seem to affect the success of the PBL approach as much as how the teacher organized the room and the learning activities. Having adequate room to conduct science inquiries was, however, an asset in terms of convenience and efficiency.

During a typical PBL science class, students are engaged in a variety of activities. With proper classroom management and organization, the time students spend on the intended learning activities can be maximized. From the perspective of our observations, successful teachers tended to have their materials and equipment set out prior to the activity; they spent their time managing on-task behavior and assisting students in thought provoking reflection. An organized class structure that included a prelab and post-lab discussion helped students to focus efforts and provided time to reflect on what students had learned. Good questioning by the teacher helped expose any misconceptions student may have developed, allowing the teacher to help students make corrections in their thought processes. While technology was often extremely important to the learning activity, a well-organized and experienced teacher often made the difference in whether the students benefited from having the technology made available to them.

DISCUSSION OF FACTORS IMPORTANT TO THE INTEGRATION OF TECHNOLOGY IN SCIENCE INSTRUCTION

This section provides a synthesis of the ideas presented in the previous section. Factors important to the process of integrating technology in an inquiry-based science classroom are discussed. These include (1) the concept of highly qualified and highly effective science teachers, (2) the selection of PBL questions and technology integration, (3) the need for technology training and technical support, (4) motivation and technology use, and 5) the effective integration of technology to promote learning.

Highly Qualified and Highly Effective Science Teachers

One important characteristic of science teachers who are considered to be highly qualified is that they have a good understanding of the subject they teach. A science teacher's competency is therefore established through evidence of subject-matter knowledge. In other words, science teachers must have a good understanding of the facts, laws, principles, and theories of science they will be expected to teach their students. In science, however, being highly qualified also means that teachers understand that science is not only a body of knowledge that needs to be learned but also a way of thinking and investigating that involves the collection and analysis of empirical evidence. Therefore a highly qualified science teacher will also have a thorough understanding of the procedural knowledge and process skills required to think and investigate scientifically. Science instruction, and associated learning goals, should reflect this multifaceted vision of science (NRC, 1996).

Being highly qualified is not the same as being highly effective. It is clear that knowing about a topic and being able to inspire student learning are very different competencies. Effective science teachers not only know and understand the true nature of science, they are able to design their science instruction in such a way that students develop science process skills as they gain science content knowledge. The best science teachers are passionate about teaching science and are often curious and inquisitive, encouraging their students to be the same. They are well organized and have good classroom management skills. They also implement technology in ways that improve the potential for learning goals to be accomplished with an understanding that technology facilitates different learning objectives in different ways.

PBL Question Selection and Technology Use

While an instructional approach like PBL has the potential to facilitate learning goals related both to the building of science process skills and to learning specific science content material, the specific PBL question selected for study often determines whether the intended learning goals will be achieved. Research questions (i.e., questions that can be answered through information gathering) tend to facilitate learning goals associated with gaining content knowledge. These types of questions lend themselves to technology that facilitates information searches (e.g., computers and the Internet, information databases, and word-processing software). Testable questions are questions answered through experimentation. In addition to requiring students to think scientifically, testable questions tend to facilitate the acquisition and practice of science process skills. Often technology with a unique or specialized purpose and function are best for answering this type of question.

Effective science teachers understand that the type of questions selected for a specific activity will determine which technology is appropriate to use and how that technology might best be utilized. When the instructional task is well-structured, technology can automate some of the time-consuming aspects of the activity and help focus student learning on the intended learning goals. In order for this to happen in a PBL unit, a teacher must recognize the benefits of having students answer each type of question and guide students' selection of a PBL question that lends itself to the specific learning goals of the course. Whether the selected questions and technology used will help students develop the desired content knowledge is fundamentally dependant on whether the question is aligned with established state learning standards. Having the teacher select a "big idea" from the content standards to drive the selection of a testable question allows students to learn what they will be tested on as they develop important process skills (Rezba, Sprague, McDonnough, & Matkins, 2007). The intended learning objective should be clearly understood and should drive the question selection, which will then determine the best way technology can be used. In other words, the technology should not drive the instruction; rather the course learning goals should drive the technology integration.

Technology Training

In order for technology to be used effectively as a learning tool, both teachers and students must first become familiar with its purpose and operation. Being able to use the technology is an essential prerequisite to the effective utilization of technology as a learning tool. For both teachers and students there is a learning curve associated with using technology that is new to them; using and practicing with the equipment reduces frustration and problems when students are expected to complete a learning activity enabled by the technology.

Teachers who take the time and effort to become proficient with the technology are often more efficient and more thorough at helping students learn to use the equipment. The self-discovery method of allowing students to "play" with the technology in order to become acquainted with its function and utility, while common, is often an ineffective training technique. A self-discovery training method alone often fails to ensure that students understand the full functionality of the equipment and that they are able to operate the equipment in an effective manner. Providing guided practice with the equipment tends to reduce the amount of time required to help students become familiar with equipment and allows students as a group to get started on their primary learning activities more quickly. In terms of instructional effectiveness and efficiency, a guided practice approach seemed to be much better than a self-discovery approach. Cookbook type experiments are appropriate for students to learn how to use equipment and practice implementing a scientific inquiry (Rezba, Sprague, McDonnough, & Matkins, 2007); however, students will eventually need to become familiar enough with the equipment to use it without step-by-step instruction.

Technical Support

It is inevitable that technology will at times fail. When this happens, even the most technologically savvy teacher needs to have technical support. When technology does not function as intended, a lack of technical assistance results in frustration and often times an inability to complete the planned learning activity. A good technology integration plan will include funds for technical support and funds to repair or replace faulty and broken equipment.

Motivation to Use Technology Versus Using Technology to Learn

It is commonly believed that learning can be enhanced through the use of technology (U.S. Department of Education, 2002); however, learning does not take place simply because technology is used. Students are obviously enthusiastic about using technology; yet, students must go beyond seeing technology as an entertaining toy and begin seeing the technology as a tool to accomplish specific learning activities. Teachers sometimes mistake interest in the technology and activity involving technology for learning. Motivation to use technology is not enough; students must get past the novelty of the technology and begin to use it because they see how the technology, as a tool, will facilitate their learning. When this happens the technology becomes transparent, almost invisible to the learning process. Equipment with which students are familiar is more likely to be used as a learning tool. Once students start focusing on the task of completing an assignment, using the technology simply becomes a way to accomplish the learning tasks assigned to them.

Constructing knowledge is a human activity that can be facilitated by technology; yet, even when technology is used effectively as an educational tool, technology will not correct misconceptions that occur as the result of faulty logic, erroneous interpretations, or the lack of experience or knowledge. An effective science teacher will make sure that students carefully reflect. He or she will take time to help students correct any misconceptions that occur as they build understanding and science knowledge.

Using Technology in the Science Laboratory Efficiently

Learning takes time, and technology can help educators make the most of their instructional minutes. In the science laboratory, a carefully selected testable question with the assistance of appropriate technologies can often help focus students' attention to higher order learning objectives like analysis of results. For example, electronic probes coupled with computer software can facilitate gathering large amounts of data in a short amount of time. The collection of real time data can quickly be organized and displayed in graphs and tables, leaving time for students to focus on the task of understanding the data, interpret the results, and relate results to what is known about the topic, associated vocabulary, terms, definitions, and theories. This will only happen if the learning activity is well-structured and the equipment and materials are well-managed, allowing students to actually get to the intended learning task.

There are many factors that will affect the successful and efficient integration of technology into a science classroom. Among the most important factors are the classroom management and organizational skills of the teacher. Creating a laboratory setting, even if it is just with tables or clustered desks, helps promote a collaborative setting where students communicate with one another by sharing ideas, perspectives, and materials. These are behaviors students may not be skilled at initially but will improve with practice and guidance from the teacher. Having the required technology and materials prepared and organized before hand helps ensure students will not waste time looking for required equipment. Structuring the lesson so there is a prelab at the beginning and post-lab at the end, perhaps 5 to 10 minutes for each, sets the stage for the investigation and brings students' attention to the learning task they are expected to complete. During the pre- and postlab sessions the teacher may review the material and technology to be used, discuss key concepts and vocabulary; share previous findings; focus students attention to safety issues, examine measuring techniques and procedures; frame the investigation within a "big picture" view of the science concept they are to learn; and question students to check accountability and proper concept formation. During lab time the teacher moves from group to group, checking student progress, assisting with the technology, promoting students' use of reasoning skills, and encouraging them to explain why and what is happening in their investigation. This is not something every teacher will do in exactly the same way, but it is something that needs to be done if the technology is to be used effectively and efficiently to help facilitate the learning goals of the course.

CONCLUSIONS

Several patterns and themes emerged through analysis of the evaluation results. Clearly the simple goal of using technology seems somewhat unimportant by itself. There are several factors that will likely affect how effective technology integration will be at improving student learning in an inquiry-based science classroom. Table 8.1 provides an overview of these factors.

Table 8.1. Factors Important to Technology Integration in Inquiry-Based Science Instruction

Highly Qualified and Effective Teachers

- have a clear understanding of the true nature of science, which includes both content knowledge and process skills.
- model the inquiry process in learning the science content, processes, and technology their students will use.
- understand that educational technology includes any tool, electronic or mechanical, which can be used to help students accomplish specified learning goals.
- have a working knowledge of the function and utility of the technology they employ.
- align technology use to accomplish specific learning objectives in the course and not just for motivational purposes.
- apply a guided practice approach when teaching students the function and operation of equipment.
- encourage students to choose the technology best suited for conducting their own investigations.
- understand that technology can enable and automate learning activities, but it is the teacher not the technology that helps students correct any erroneous conclusions.
- organize the learning activity, managing the materials and technology properly in order to increase the likelihood that students will accomplish the intended learning objectives.

Using technology can improve the effectiveness and efficiency of science instruction, but only when it is integrated properly into the instruction. Technology integration is done best when the science teacher has a clear understanding of the true nature of science (i.e., both content knowledge and process skills), developed a working knowledge of the function and utility of the technology, and aligns technology use with the learning objectives of the course.

The intended learning goals need to be clearly understood and should drive the questions students attempt to answer in class. The specific question selected for the learning activity will determine the best way technology can be used as well as which technology should be used. Selecting a "big idea" from the content standards to drive the selection of a testable question often allows students to learn science content they will be tested on in state standardized assessments and simultaneously develop important process skills (Rezba, Sprague, McDonnough, & Matkins, 2007).

Both teacher and students need to be trained in the use of the technology because technology is more likely to be used effectively when teachers and students are familiar with the equipment. In terms of instructional efficiency and effectiveness, a guided practice approach to teaching students the function and operation of equipment seems to be more effective than a self-discovery approach. While technology is often in and of itself enticing to students, technology tends to be most useful once the equipments' novelty fades and students figure out how the equipment can be used to accomplish specific educational tasks. When this happens the technology becomes transparent to the learning process. Students who are intent on learning will use technology not only because the technology is fun, but because they see it as a tool to facilitate their inquiries and assist them in completing the learning tasks assigned to them. Students whose motivation is based primarily on using the equipment often tend to lose interest in the learning task once the equipment is no longer needed or in situations where they are not the ones operating the equipment.

The technology-enhanced PBL inquiry model does tend to facilitate students' inquiry and motivate students to participate in learning activities. At times, however, an individual student's lack of foundational knowledge and experience can produce misconceptions about the science concepts they are being taught. Technology can support scientific inquiry and analysis by enabling and automating some of the time-consuming aspects of the prescribed learning activities, thus helping to focus student learning on specific learning objectives. It cannot, however, replace intelligent human guidance as students build understanding and knowledge. Technology will not correct any misunderstandings that result from faulty logic or misinterpretation of the data. The teacher might rely on technology to make a learning activity more effective, but the teacher, not the technology, will need to help students correct any misconceptions and erroneous conclusions when they occur.

Technology can improve science instruction in a laboratory situation but only when the learning activity is well-structured and the equipment and materials to be used are well-managed, allowing students to actually get to the intended learning task. The successful and efficient integration of technology into a science classroom often depends on the teacher's ability to organize the activity properly. The teacher must organize and structure the lab activity taking into account the limitations of the physical facilities, the equipment and materials needed, the amount of time available, and the students' needs to be guided in their exploration of the topic. And while there are no guarantees that the use of technology will result in student learning, proper integration of technology can increase the likelihood that students will learn in the science classroom.

REFERENCES

Barrows, H. S. (1996). Problem-based learning in medicine and beyond: A brief overview. In L. Wilkerson & W. H. Gijselaers (Eds.), *Bringing problem-based learning to higher education: Theory and practice* (pp. 4-12). San Francisco: Jossey-Bass.

Barrows, H. S., & Tamblyn, R. S. (1980). *Problem-based learning and approach to medical education*. New York: Springer.

Bray, B. (2005). Working with reluctant teachers. *Technology & Learning, 25*(11), 12.

Chiapetta, E. L., & Koballa, T. R., Jr. (2002). *Science instruction in the middle and secondary schools* (5th ed.). Upper Saddle River, NJ: Prentice-Hall.

Chung, J., & Chow, S. (2004). Promoting student learning through a student-centered problem-based learning subject curriculum. *Innovations in Education and Teaching International, 41*(2), 157-168.

Crockett, C. (2004) What do kids know—and misunderstand—about science? *Educational Leadership, 61*(5), 34-37.

Darling-Hammond, L. (2000). Teacher quality and student achievement: A review of state policy evidence. *Educational Policy Analysis Archives, 8*(1). Retrieved March 30, 2007, from http://epaa.asu.edu/epaa/v8n1

Davies, R. (2006). Learner intent and online learning. In C.S. Sunal, E. K. Wilson, & V. H. Wright (Eds.), *Perspectives on distance learning series: Research on enhancing the interactivity of online learning* (pp. 5-26). Greenwich, CT: Information Age.

Falk, J., & Drayton, B. (2004). State Testing and Inquiry-Based Science: Are they complementary or competing reforms? *Journal of Educational Change, 5*(4), 345-387.

Fisher, C., Dwyet, D., & Yocam, K. (1996). *Education and technology: Reflections on computing in classrooms.* San Francisco: Jossey-Bass.

Glazer, E., Hannafin, M., & Song, L. (2005). Promoting technology integration through collaborative apprenticeship. *Educational Technology Research & Development, 53*(4), 57-67.

Goldhaber, D. D., & Brewer, D. J. (1996). *Evaluating the effect of teacher degree level on educational performance.* East Lansing, MI: National Center for Research on Teacher Learning. (ERIC Document Reproduction Service No. ED406400)

Gross, P. R. (2005). *Less than proficient: A review of the draft science framework for the 2009 national assessment of educational progress.* Retrieved March 14, 2007, from http://www.edexcellence.net/doc/LessThanProficient.pdf

Harmon, M. (1995). The changing role of assessment in evaluating science education reform. *New Directions for Program Evaluation, 65*, 31-51.

Harwood, W. (2004). An activity model for scientific inquiry. *Science Teacher, 71*(1), 44-46.

Hembree, R., & Dessart, D. (1986). Effects of hand-held calculators in precollege mathematics education: A meta-analysis. *Journal for Research in Mathematics Education, 17*(2), 83-99.

Interstate New Teacher Assessment and Support Consortium. (2002). *Model standards in science for beginning teacher licensing and development: A resource for state dialogue.* Retrieved March 30, 2007, from http://www.ccsso.org/content/pdfs/ScienceStandards.pdf

Jones, B., Valdez, G., Nowakowski, J., & Rasmussen, C. (1994). *Designing learning and technology for educational reform.* Oak Brook, IL: North Central Regional Educational Laboratory.

Kleiman, G. M. (2000). Myths and realities about technology in K-12 schools. In the Harvard Education Letter Report [online]. *The digital classroom: How technology is changing the way we teach and learn.* Retrieved April 15, 2007, from http://www.edletter.org/dc/kleiman.htm

Kozma, R. B., & Croninger, R. G. (1992). Technology and the fate of at-risk students. *Education and Urban Society, 24*(4), 440-453.

Lee, H. S., & Songer, N. B. (2003). Making authentic science accessible to students. *International Journal of Science Education, 25*(8), 923-948.

Lemke, C. (2005). Measuring progress with technology in schools. *THE Journal, 32*(9), 16-20.

Linn, R. L., & Miller, D. M. (2005). *Measurement & assessment in teaching* (9th ed.). Saddle River, NJ: Prentice-Hall.

Mendro, R. L. (1998). Research findings from the Tennessee Value-Added Assessment System (TVAAS) database: Implications for educational evaluation and research. *Journal of Personal Evaluation in Education, 12*, 247–256.

National Research Council (1996). *National science education standards.* Washington, DC: National Academy Press.

National Science Teachers Association. (2007). *The integral role of laboratory investigations in science instruction* (Position paper). Retrieved April 15, 2007, from http://www.nsta.org/positionstatement&psid=16

O'Neil, T., Sireci, S., & Huff, K. (2004). Evaluating the consistency of test content across two successive administrations of a state-mandated science assessment. *Educational Assessment, 9*, 129-151.

Peck, C., Cuban, L., & Kirkpatrick, H. (2003). High-tech's high hopes meet student realities. *Education Digest, 67*(8), 47-54.

Popham, W. J. (1999). Why standardized tests don't measure educational quality. *Educational Leadership, 56*, 8-15.

Rezba, R. J., Sprague, C. R., McDonnough, J. T., & Matkins, J. J. (2007). *Learning and assessing science process skills* (5th ed.). Dubuque, IA: Kendall/Hunt.

Shockley J. S., McGurn, W., Gunning, C., Graveley, E., & Tillotson, D. (1989). Effects of calculator use on arithmetic and conceptual skills of nursing students. *Journal of Nursing Education, 28*(9), 402-405.

Stedman, L. C. (1997). International achievement differences: An assessment of a new perspective. *Educational Researcher, 26*, 4–15.

Stucker, H. (2005). Digital "natives" are growing restless. *School Library Journal, 51*(6), 9-10.

Torp, L., & Sage, S. (2002). *Problems as possibilities: Problem-based learning for K-16 education* (2nd ed.). Alexandria, VA: Association for Supervision and Curriculum Development.

U.S. Department of Education. (2000). *Before it's too late: A report to the nation from the national commission on mathematics and science teaching in the 21st century.* Retrieved March 15, 2007, from http://www.ed.gov/inits/Math/glenn/report/pdf

U.S. Department of Education, Office of Elementary and Secondary Education (2002). *No Child Left Behind: A desktop reference.* Retrieved March 15, 2006, from https://www.ed.gov/admins/lead/account/nclbreference/page.html

U.S. Department of Education. (2003). *Federal funding for educational technology and how it is used in the classroom: A summary of findings from the integrated studies of educational technology.* Retrieved March 15, 2006, from https://www.ed.gov/about/offices/list/os/technology/evaluation.html

Weston, T. (2005). Why faculty did—and did not—integrate instructional software in their undergraduate classrooms. *Innovative Higher Education, 30*(2), 99-115.

White, B. Y., & Frederiksen, J. R. (1998). Inquiry, modeling, and metacognition: Making science accessible to all students. *Cognition and Instruction, 16*(1), 3-118.

CHAPTER 9

IMPACT OF TECHNOLOGY ON INFORMAL SCIENCE LEARNING

David A. Ucko and Kirsten M. Ellenbogen

Diverse forms of educational technology facilitate and serve as resources for informal learning in a variety of out-of-school settings, including science museums and the home. This chapter examines studies of technology-based exhibits, mobile technology, and Web sites, emphasizing those experiences designed for learning within the science museum context. It also addresses studies of narrative media, such as television and large-format film. Studies in these areas face greater challenges than those carried out in the classroom, and they are relatively limited in number. Nevertheless, they indicate the potential of educational technology in these contexts, which emphasize a learner focus, the role of motivation, social interaction, and just-in-time learning.

INTRODUCTION

The intent of this chapter is to provide an overview of educational technologies used for informal learning in science, technology, engineering, and math (STEM) in out-of-school settings. Because readers of this volume may be less familiar with this context, the chapter begins with a brief

The Impact of the Laboratory and Technology on Learning and Teaching Science K-16, pp. 239–266

overview of informal STEM learning, followed by discussions of the various roles played by technology. The chapter concludes with a review of issues raised by the use of technology for informal learning, and directions for further study.

Informal Science Learning Environments

Rather than being guided by a teacher and a standards-based curriculum, informal learning is voluntary and self-directed, driven by intrinsic motivation based on widely varying personal interests. Informal learning is episodic, typically involving activities of relatively short duration, often mediated within a family or other social group. It can be considered the primary way in which most people learn most of the time, since the hours spent in school represent less than 10% of the average person's life (Jackson, 1968; Sosniak, 2001; U.S. Department of Education, 1991).

In content areas related to STEM, informal learning is facilitated by institutions established for that purpose, notably science-technology centers or science museums broadly defined (e.g., science centers, natural history museums, planetariums, zoos, aquariums, botanical gardens, arboreta, nature centers, and science-rich children's museums). They provide rich environments that offer visitors the impact and multidimensionality of tangible objects, multisensory experiences, and reality-based learning situations (Sweeney & Lynds, 2001), along with opportunities for social interaction. Some 75 million children and adults visit more than 300 U.S. science museums annually, primarily as family groups (Association of Science-Technology Centers [ASTC], 2005). In addition to these institutions, libraries, community-based organizations, and diverse forms of media, including the Web, provide informal education in aspects of STEM along with other content areas to form a loosely interconnected informal science learning infrastructure (Lewenstein, 2001; St. John & Perry, 1996). Recent technology-based vehicles for learning enable organizations within this infrastructure to design virtual experiences that can overcome geographic, physical, and time constraints, enabling these institutions to reach an even greater portion of the public.

In many respects, the term "informal" is unfortunate because it is based on a contrast with "formal" classroom learning. This distinction is fuzzy, somewhat arbitrary, and at times incorrect; exceptions are common. For example, teachers in many classrooms employ the hands-on activities typical of informal learning, and formal instruction often takes place within informal learning institutions that have hidden curricula (Vallance,

1995). Nevertheless, at least in the case of museums, informal education can be differentiated in general from formal education by the nature of the stimuli, physical environment, overt behaviors, social contacts, and learning consequences (Bitgood, 2002). As encapsulated in the often-repeated quote by the former physics professor and teacher who founded the Exploratorium, "no-one ever flunks a museum" (Oppenheimer, 1975, p. 11).

The term "informal" has been applied to indicate the setting (outside of school), the process (without an instructor or self-directed), and the audience (other than students). In response to this lack of specificity, some prefer to use the term "free-choice learning," emphasizing a characteristic of the learning (Falk & Dierking, 2002, p. 6). Others create a distinction between informal and non-formal. Also defined in various ways, "non-formal" is often associated with organized learning outside the formal educational system (Carlson & Maxa, 1997). Related terms include lifelong learning (Chapman & Aspin, 2000), out-of-school learning (Rennie, Feher, Dierking, & Falk, 2003), recreational learning (Ucko, 1998), and experiential (Kolb, Boyatzis, & Mainemelis, 2000) or experience-based learning (Andresen, Boud, & Cohen, 2000). Some authors make distinctions between learning that is intentional versus learning that occurs "in the wild" in everyday settings, which may be considered accidental, incidental, or implicit (Bransford et al., 2006). The Learning in Informal and Formal Environments (LIFE) Science of Learning Center, for example, places emphasis on "emergent occasions of learning that occur in homes, on playgrounds, among peers, and in other situations where a planned educational agenda is not authoritatively sustained over time" (Bransford et al., 2006, p. 216). Intentional and incidental learning can reinforce one another, however, as in the coconstruction of "islands of expertise" through family activities that build on individual interests over time (Crowley & Jacobs, 2002, p. 333).

For the purposes of this chapter, "informal" learning will be broadly considered to encompass learning experiences facilitated by science museums and similar institutions, as well as other forms of self-directed learning that occur outside of school, such as in the home. The focus will be on those experiences that have been specifically designed for informal learners.

Studies on Informal Learning

In addition to the difficulties of sharply defining its domain, informal learning outside the classroom is typically more challenging to study than formal. The number of independent variables is considerably greater as a result of the many factors that influence an individual's personal, sociocultural, and physical context (Falk & Dierking, 2000). As an example, preliminary studies within the museum setting indicate that certain variables—prior visitor knowledge, motivation, expectations, within group social interaction, advance organizers, and exhibition design—appear to play the most significant roles; other factors that influence museum learning include prior visitor experience, prior interest, choice and control, between group social interaction, orientation, and architecture (Falk & Storksdieck, 2005). Even focusing on a single concern such as making appropriate use of learning progressions, an area of active research for developing classroom science curricula (Grandy & Duschl, 2005), is far more daunting when applied to heterogeneous audiences of informal learners.

The scope of learning outcomes is considerably broader than the conceptual learning and cognitive gains traditionally assessed in the classroom, an emphasis accentuated by mandatory large-scale testing. Informal learning may have its greatest impact on attitudinal or behavioral change and the affective domain, stimulating an individual's interest in additional exploration and further learning (Meredith, Fortner, & Mullins, 1997). Yet affective state information is notably much more difficult to measure than cognition (Picard et al., 2004).

In addition, there are methodological obstacles to conducting research on "non-captive" audiences, whether museum visitors, television viewers, or Web users. Studies often focus disproportionately on concerns related to usability, such as navigation. This emphasis contributes to the ease with which users can access learning resources, but it obscures larger, more critical issues. For example, understanding how, why, and to what end people use science museum Web sites would help designers better select, organize, and present learning resources and activities. Understanding the impact of such experiences can also provide insight into how best to position this virtual resource in relationship to the physical science museum, other museum Web sites, and complementary aspects of the learning infrastructure (e.g., books, magazines, television).

For these and other reasons, research and evaluation studies on informal learning are far fewer than those conducted in the classroom, and the fields are less well established. However, this situation is improving. Evaluations of informal learning practice, more common than research per se (although the two are sometimes conflated), are being carried out in greater number. The requirement that all projects funded by the National Science Foundation's (NSF) Informal Science Education (ISE) program include at least formative and summative evaluations (NSF, 2006) has been a driver over the past decade; most projects also involve front-end studies of target audiences and may include remedial evaluations as well. Beginning in 2004, the ISE program solicitation explicitly asked proposers to demonstrate how projects build on the results of prior work. To facilitate that outcome and enhance knowledge accumulation, the program requires grantees to postsummative evaluations at the Web site www.informalscience.org and has funded the Association of Science-Technology Centers (ASTC) to develop an ISE Resource Center (DRL-0638981) to disseminate research and practice further at the Web site www.insci.org.

Research and evaluation efforts have been supported by an increasing professional presence in the field. The Visitor Studies Association (www.visitorstudies.org) was founded in 1988, and the American Association of Museums formally established the Committee on Audience Research and Evaluation (CARE) 2 years later. The National Association for Research in Science Teaching has had an informal science learning strand since 1995. The Informal Learning Environments Special Interest Group (ILER-SIG) was established within the American Education Research Association in 1997. That same year, the first annual international conference on Museums and the Web (www.archimuse.com/conferences/mw.html) was held.

Although published informal learning research still represents a small niche within the literature of educational research, especially in the museum setting, these studies have increased in number over the past decade (Dierking, Ellenbogen, & Falk, 2004). The growth is reflected in the increasing frequency of special issues of educational research journals dedicated to informal science education in the last 15 years. They include the *International Journal of Science Education* (1991), *Science Education* (1997), *Journal of Research in Science Teaching* (2003), and most recently a supplemental issue of *Science Education* (2004) published as part of an NSF-funded initiative designed to coalesce the last decade of research on learning in museums into frameworks for practitioners.

The increase of professional organizations and publication outlets indicate the growth of a field that historically has been characterized by extreme diversity in educational training and professional affiliations. As a result, research findings have been splintered into a wide array of publications or left unpublished. Consequently, in the view of one observer, "the result is a set of individual research case studies, which are difficult to compare and from which it is almost impossible to generalize" (Falk, Dierking, & Storksdieck, 2005, p. 5). The National Research Council's Board on Science Education received NSF funding (DRL-0545947) to conduct a consensus study that will attempt to draw together the disparate informal science literatures, synthesize the state of knowledge, and articulate a common framework for the next generation of research on informal STEM learning. This chapter will employ the lens of educational technology to help bring focus to a portion of the research literature.

ROLE OF TECHNOLOGY IN INFORMAL LEARNING

As a result of developments in technology and their applications in museums, the home, and other settings over the past decade, technology-based proposals have been the fastest growing category of proposals to NSF's ISE program relative to proposals for museum/exhibit, television/radio, and community/youth projects. The field of informal STEM education has embraced technology for a variety of reasons. One strong impetus has been the desire to engage visitors through "hands-on" experiences, a hallmark of science museums today. Such activities are consistent with inquiry-based learning (Olson & Loucks-Horsley, 2000), constructivist (e.g., Anderson, Lucas & Ginns, 2003; Hein, 1995), and sociocultural learning theories (e.g., Ellenbogen, 2003; Leinhardt & Knutson, 2004), which provide underlying educational frameworks for learner-centered engagement with exhibits and related interactive activities. Another derives from the necessity to offer enjoyable learning experiences that cannot be readily duplicated at home or in school. This need is especially great for science museums, which must continually attract new and repeat visitors to their facilities to generate revenue.

Not surprisingly, there is overlap in the use of certain educational technologies within and outside the classroom, such as computer- and video-based programming. This chapter will emphasize studies based on those technologies and applications most commonly employed in informal settings in a way that is distinct from their classroom-based applications. They can be sorted more or less within existing taxonomies for educational technology, including those developed primarily to encompass formal learning applications. For example, most fit within the categories of "Learning Tools" [II.A.2 Educational Media and II.6 Interactive Learning Media] (Educational Technology Research & Assessment Cooperative [ETRAC], 2002) or correspond to "Media for Inquiry" (Bruce & Levin, 1997, p. 6). More closely aligned is the taxonomy of digital technologies developed for museum learning opportunities (Hawkey, 2004), which separates those offered on-site and online. Although these taxonomies provide overall frameworks, they are necessarily arbitrary and provide a classification system based on a particular set of criteria; alternative classification schemes could be derived from the learning environment, degree of interaction, or other aspects of the technology. For the purposes of this chapter, technology in informal science learning environments will be discussed in the following categories: technology-based exhibits, mobile technology, Web sites, and narrative media, such as television and large format films. Science museum applications will be emphasized over other settings because they have received more attention, especially compared to research on educational technology in the home (Kafai, Fishman, Bruckman, & Rockman, 2002).

Technology-Based Exhibits

Permanent and traveling exhibitions are the primary means by which most science museums engage their visitors with aspects of STEM content. These exhibitions contain exhibit elements, or exhibits, that typically involve technology-based interactive devices or "interactives," which may be mechanical, electronic, or multimedia-based. They can range from simple low-tech devices, such as direct manipulation of a simple phenomenon, to those that are complex and high-tech, such as scientific instrumentation. Historically, science museums have pioneered the use of technology, such as computers, for public use. As an example, a major permanent exhibition was developed in the early 1980s using then state-of-the-art Texas Instrument TI-99/4A personal computers on which visitors registered views on the impact of technology on their lives (Ucko, 1983).

Interactive exhibits offer visitors the opportunity to explore real (and sometimes simulated) scientific phenomena, as well as aspects of historic and state-of-the-art technology. Interactives based on classical physics tend to be the most widespread because the phenomena lend themselves readily to direct visitor manipulation, although exhibits based on biology (e.g., Colson, 2005) and chemistry (e.g., Ucko, Schreiner, & Shakhashiri, 1986) also have been developed. Because the term "interactive" encompasses an extremely wide range of experiences, from simple and individual to complex and collaborative (Heath & vom Lehn, 2003), it is challenging to generalize findings about learning derived from use of these devices. In any event, one would do well to keep in mind, as these authors point out, that it is the learner who is interactive.

Science museums have begun to explore the use of newer technologies to create augmented and virtual environments (Roussou, 2000). Research on their impact, like other areas of technology in informal learning environments, is dominated by usability studies. An exception is a series of studies by Roussou and her colleagues (Roussou et al., 1999) that has demonstrated that collaboration in virtual reality (VR) settings, where users interact with a computer-simulated environment, increases the learning experience; much of this work, however, was conducted in the laboratory or in classrooms. More recently her work has demonstrated the critical role of interactivity for learning in VR experiences in museums and other informal learning environments (Roussou, 2004). The worlds of VR and augmented reality pose particular problems for integrating interactivity. How can people who visit a museum in a group together share the same VR experience? What distinctions are made between mere navigational interaction and control over the VR environment? These two problems weaken many of the existing efforts to measure learning in VR environments. Participants report a high level of engagement and enjoyment, but few qualitative or quantitative measures have demonstrated conceptual learning. In a study on the effects of learning at a virtual reality exhibit, de Strulle (2004) discusses an exhibit developed by the Reuben H. Fleet Science Center in San Diego with multiple kiosks that enable learners to enter a shared VR experience as personalized avatars. Within this interactive virtual environment, de Strulle found an array of critical design and instructional issues shown to facilitate and detract from learning in virtual spaces with implications for the development of future VR exhibits. Further work will be needed to demonstrate that this type of interactivity effectively mediates learning along with serving as an attractor.

Technology holds great potential to support inquiry practices in museums (Ansbacher, 1997). It has proven to be an effective scaffolding tool that helps learners engage in domain-specific inquiry (Lin, Davis, & Bell, 2004). Embedding supports for discipline-specific thinking into object- and phenomenon-based experiences changes learners' experiences by allowing them to participate in disciplined inquiry. Technology-based tools can supplement and assist, but do not necessarily have to replace, direct experiences in a museum. Rather, they make objects and phenomenon a central component of an inquiry, and then can further support learners as they extend their investigations beyond simply reading exhibit labels or even beyond a one-time visit.

Visitors tend to use technology-based exhibits more frequently, for a longer period of time (Serrell & Raphling, 1992). One of the explanations for this longer hold time, which is a measure of visitor attention, is "technological novelty" (Sandifer, 2003). In this study, technological novelty was defined when an exhibit either contains visible state-of-the-art devices or illustrates, through the use of technology, phenomena that would otherwise be impossible or laborious for visitors to explore on their own.

A concern sometimes raised is that technology-based exhibits may reduce visitors' interactions with other exhibits or objects in the museum, or worse, replace authentic experiences. In the case of object-based exhibits, technology appears not to compete but to engage users in a different kind of learning (Eberbach & Crowley, 2005). There are also concerns that technology may decrease social interaction that is a hallmark of informal learning experiences. The interfaces on technology-based exhibits, such as touch screens or joysticks, are often designed for one person (Flagg, 1994). Unless social interaction is prioritized in the design of technology-based exhibits, people will continue to be hampered in their efforts to use technology-based exhibits in social groups (Heath, vom Lehn, & Osborne, 2005).

Mobile Technology in Museum Exhibitions

Hand-held personal data assistants and data probes have been used to extend science learning beyond the classroom for years. These devices are ideal for just-in-time learning (Bransford, Brown, & Cocking, 2000) and field research (e.g., Soloway et al., 1999). Many of these projects occupy an overlapping space that falls between informal and formal learning environments, and hold potential for linking these environments across the educational infrastructure.

Personal or mobile devices are perhaps the most rapidly growing category of technology for informal learning environments. Mobile devices such as cellular phones, Personal Digital Assistants (PDAs), and portable digital media players (e.g., iPod) offer the potential to support self-directed and customized learning anytime and anyplace. An example of a project underway is "Science Now, Science Everywhere" (DRL-0610352) at the Liberty Science Center in New Jersey, where mobile phones are being used to expand exhibit learning. Podcasts and vodcasts (video podcasts) are becoming increasingly available on Web sites of educational radio and television and programs. As noted in a recent review (Scanlon, Jones, & Waycott, 2005), emphasis should be placed on the mobility of the learner rather than the specific nature of the mobile technology, whose unique affordances include accessibility and immediacy, in addition to portability.

Museums have been relative early adopters of mobile technology in an effort to customize or to supplement existing exhibit labels and interactive components. Audio tour technology has long been used to supplement and enhance exhibitions. Mobile technology can now serve as highly personalized guides. For example, the Exploratorium, in partnership with Hewlett-Packard Laboratories and The Concord Consortium, developed the Electronic Guidebook, a portable device that allows the museum to communicate with its visitors in a customized manner (Fleck et al., 2002a; Hsi, 2003). Visitors are given a PDA, which serves as a tool for informing, suggesting, and remembering that they can carry with them in the museum. These Electronic Guidebooks provide information by creating what is essentially a homepage for each exhibit that can be accessed on the PDA; they also function as electronic scrapbooks by allowing visitors to bookmark Web pages. Findings led the researchers to focus on more streamlined experiences. Fleck and colleagues (2002b) tested a simpler version that focused only on the remembering or scrapbook function. Following up several weeks later, the researchers found that most of the test subjects revisited the Web pages that were bookmarked during their museum visit.

In addition to this kind of handheld device, Radio Frequency Identification (RFID) tags or transponders, small and inexpensive enough to be integrated into almost anything, can be used to personalize the visitor experience. In the NetWorld exhibition at the Museum of Science and Industry, Chicago, for example, RFID technology offers a central means of interacting with the exhibits by allowing visitors to create personal avatars. At The Tech Museum of Innovation in San Jose, CA, visitors to Genetics: Technology With a Twist can grow bacteria in a wet lab, using RFID-based Tech Tags to post results at their very own Web site, allowing them to track their experiment's progress online days later. A study showed that less than half intended to follow up and see the customized Web page (Eberbach, 2006). These findings contrast with those of a national study of museum Web site users that found that more than half were likely or very likely to visit a museum's Web site after visiting the physical location (Marty, 2005), raising questions for further research on these applications of technology.

Use of mobile technology in museums presents great challenges. In the Tech Tags study, visitors reacted very positively to the concept of being able to use technology to customize their museum visit, but at the same time, noted frequent technical complaints. The discrepancy between ideal and practical uses of mobile and ubiquitous computing will need to be addressed directly as they become more common. In an institution filled with hands-on exhibits, the devices compete for attention, along with the museum visitor's hands, eyes, and ears. Thus, there is a danger of the mobile device displacing the exhibit (vom Lehn, Heath, & Hindmarsh, 2005). In addition, the device typically prioritizes the individual user over the group. To overcome this limitation on social interaction, visitors sometimes group together in an attempt to synchronize experiences, pressing the "start" button at the same time or connecting multiple headphones to the same device. They may also divide tasks and share information, with one person attending to the mobile device and the other attending to the exhibit.

The challenges of integrating mobile technology into the museum experience are being addressed in numerous projects that are keeping pace with new developments in hardware and software (e.g., cell phone capabilities), as well as new uses for technology (e.g., podcasting). Less is known, however, about the learning impacts of these devices. One proposed rubric for analyzing the impact of handheld technologies is the extent to which they embody the characteristics of portable versus static and personal versus shared (Naismith, Lonsdale, Vavaoula, & Sharples, 2005). Sharples (2003) proposes that the most important requirements for successful mobile technology are portability, adaptability, availability, and usability. The focus, for now, remains on usability as museums explore not only new technology, but also metaphors to help visitors understand the function of the technology, privacy statements to assuage concerns, and even public orientation and training sessions (Hsi & Fait, 2005).

Museum Web Sites and Web-Based Museum Experiences

Web sites have become a well-established technology used to supplement and scaffold museum exhibits. Science museums employ Web sites as an opportunity to expand their audience and activities by providing not only highlights of their exhibits or programs, but also synergistic content and programming. The Internet offers tremendous potential for self-directed learning, whether accessed at home, in a museum, or in a library. It has been called "one of the most powerful and important self-directed learning tools in existence" (Gray, 1999, p. 120). Characteristics that promote informal learning include providing independent access to information and resources, self-directed and self-paced use, and capability to build on prior knowledge.

Museums now use the Web for such diverse purposes as webcasts, online exhibits, and virtual tours (Spadaccini, 2006). For collecting institutions, such as natural history museums, computers have provided the public with virtual access to at least portions of the artifacts held in storage. In addition, the Web provides access to educational media, games, simulations, and scientific visualization. Some museums have created a presence on the web that rivals their physical institution. The Exploratorium (www.exploratorium.edu) and Franklin Institute Science Museum (www.fi.edu) host sites that attract millions of users annually. Allison-Bunnell and Schaller (2005) argue that any movement toward a virtual museum will require the field to reconceptualize the online science exhibit experience. For example, an object-based exhibit can become part of an online reference resource, or an exhibit that demonstrates phenomena can be transformed online to explain the underlying principle.

Not surprisingly, people are able to visit museum Web sites more frequently than a physical museum. A recent national study of museum Web site users (Marty, 2005) shows that visits to science museum Web sites are more frequent (weekly vs. quarterly), but significantly shorter than physical visits. The average museum Web site visit is generally under 12 minutes, while the average visit to science museums lasts approximately 2 hours (Semper, Wanner, Jackson, & Bazley, 2000). Half of all visitors who arrive at a home page leave after viewing only the home page, suggesting that either the person was quickly able to get the needed information or instead was uninterested, unsatisfied, or overwhelmed (Jensen, 1999).

There is limited research on the ways in which visits to museum Web sites influence or compare to visits to museums. The previously cited study of a Web site linked to exhibits through RFID technology (Eberbach, 2006) revealed that people who had access to the Web site were able to recall, on average, only one more exhibit than people who had not had access to the Web site. Preliminary research suggests that the motivations for visits to museum Web sites differ significantly from motivations for visits to physical science museums (Haley, Goldman, & Schaller, 2004). Typical motivations for science museum visits include entertainment or recreation, social activity, education, a life cycle event ("My mother always took me here, so now I take my children"), place ("We have to go to the Smithsonian while we're in Washington DC"), content interest, and practical reasons ("It's too cold to take the children to the park") (Moussouri, 1997; Rosenfeld, 1980). On the other hand, Haley, Goldman, and Schaller reviewed the research on Web site visit motivations to identify a different set of primary drivers: gathering information for an upcoming visit to the physical site, engaging in very casual browsing, and seeking specific content information as either self-motivated or assigned research.

These motivations make sense when situated in the larger context of how people in the United States use the Internet. The Pew Internet and American Life Project (Fallows, 2006) shows that 40 million Americans surf the Internet just for fun or to pass the time per day in a typical month. Surfing for fun only falls behind using e-mail, using a search engine, or getting online news. The Web has rapidly become the predominant source (52% of those surveyed) when information is sought by the public about *specific* personally-relevant science-based issues (National Science Board, 2006). These numbers reveal that the motivations of the general web user are the same as the virtual museum-goers—entertainment and seeking information.

Webcasts and virtual field trips that are broadcast online add a new dimension to museum programming. A webcast is generally a live online broadcast that allows people to participate in a tour or demonstration.

Some science museums have developed webcasts with both asynchronous and real-time communication between staff and visitors. The Darwin Centre in London, for example, offers participants the ability to communicate with staff *via* e-mail both before and during the actual webcast, and then archives past webcasts for later viewing.

Museums are also creating alternatives to museum visits through virtual field trips. Typically a museum staff member or a scientist provides a tour of a specific research site or other facility to a geographically distant school group. The program allows two-way communication so students are able to ask questions of the museum staff. In the Jason Project, which originally started in 1989 using satellite technology, students are able to observe and communicate with scientists in diverse environments, such as rainforests and wetlands. Alternatively, students visit a webcast site at the museum where they link to an otherwise inaccessible site, such as the Liberty Science Centers' *Live From: Cardiac Classroom* and the Museum of Science and Industry's *Live at the Heart* project. In these programs, students view live, open-heart surgery. They are able to examine medical instruments used during the surgery and ask questions of the medical staff.

Another interesting Web-based application is "citizen science," in which the public makes local observations that they transmit by means of the Internet to researchers who collect and analyze the resulting data. The Cornell Laboratory of Ornithology has pioneered this type of activity, which involves the public in scientific investigations (www.ebird.org); other citizen science projects involve observation of spiders, butterflies, and weather data (e.g., http://www.cocorahs.org). These projects are based on the principle that the Web allows ordinary citizens to become involved in and contribute to ongoing scientific research. Programs such as the Great Backyard Bird Count draw as many as 50,000 people submitting data during a week (Bonney, 2005). This program represents an excellent Web-based example of informal science learning that is based on three key elements: real science, real learning, and real partnerships (Barstow, 2005). These citizen science projects also engender a critical feature of Web-based informal learning experience by allowing people to interact with each other rather than an organization (Bandelli, 2005). These programs, therefore, are situated between the highly-designed and mediated experiences of science museums and the less-mediated home-based experiences.

Narrative Media

Narrative media, such as film and television, employ diverse forms of technology for informal learning experiences. Large-format film or "giant screen" theaters, located in approximately one-third of science museums as well as other venues, show science-based documentaries along with other films. (The technology typically involves 15-perforation/70 mm film projected onto a very large slightly-curved rectangular screen or a dome.) According to one survey, more than one in three Americans annually see a giant-screen film, although not all are educational (Opinion Dynamics Corporation, 2005). Because of the large frame size and extremely high resolution of the film, this technology immerses viewers into the projected image, whether photographed with special cameras or computer-generated. Planetariums employ optical or digital projection systems to create shows that incorporate images of the sky, space, and occasionally other scientific subjects. A recent article makes the case for digital full dome systems as a powerful tool for learning astronomy and calls for research studies on the best ways to use this technology (Yu, 2005). Laser projection systems, including 3-D versions, have been used in both planetarium and theater settings.

There are few studies on these technologies. Notable exceptions (e.g., Fisher, 1997) have focused on programming characteristics, such as humor, that have the potential to impact learning or impact specific audiences, such as school groups (e.g., Storksdieck, 2005). The most comprehensive study to date is a review of summative evaluations on 10 giant-screen film projects, including their supporting materials (Flagg, 2005). The evaluators typically measured changes in verbal knowledge and perceptions of scientists. All 10 of the studies showed a significant impact on viewers' verbal knowledge. Three of the 10 measured perceptions of researchers and found that half or more of their viewers felt they learned something new about the lives and work of scientists and researchers. The studies were less likely to examine the impact of the film on viewers' interests and attitudes. Only 5 of the 10 measured change in interest level, and only 2 of those 5 studies found a significant positive impact. These two films (*Stormchasers* and *Dolphins*) increased interest more interested in learning more about related topics. Less is known about changes in viewers' attitudes. A study of one film, *Tropical Rainforests*, found that adult, youth, and child viewers had a significantly more positive attitude towards rainforests after viewing it.

Shifting from museums and theaters to the home, television becomes an important potential element within the informal learning infrastructure. It is the primary source in the United States (41% of those surveyed) for *general* information about science and technology (National Science

Board, 2006). Science- and math-based television and radio programs reach some 100 million children and adults each year. Educational science programming on television, once primarily the domain of the Public Broadcasting System (PBS), can now also be found on several Discovery Channels, National Geographic Channel, The Learning Channel (TLC), NASA TV, and others. Top-rated educational programming currently includes *Zoom* (WGBH, ages 5 to 11); *Cyberchase* (WNET, ages 8 to 12); *Dragonfly TV* (TPT, ages 9 to 12); and *PEEP and the Big Wild World* (WGBH/TLC & Discovery Kids, pre-K). *NOVA* (WGBH) is the most widely-viewed science program for adult audiences. Each of these programs also offers ancillary activities on the web, making www.pbs.org one of the most popular .org sites and informal learning resources worldwide.

Science programming has been part of television from its earliest days with such shows as *Watch Mr. Wizard* in the 1950s. Shows for young audiences in particular were spurred by the strengthening in 1996 of The Children's Television Act of 1990, requiring that networks broadcast 3 hours per week of educational programming for children. Others point to the increase of the child-related economic market (Fisch, 2004a). Data on the impact of the earliest television shows is largely anecdotal (Newsom, 1952) but more recently, the study of science-related television has been extensive and more rigorous (e.g., Fisch, 2004; Flagg, 1994; Rockman et al., 1996).

What sort of learning occurs as a result of science-related television? Fisch (2004b) characterizes three critical outcomes: (1) attitudes towards science, (2) exploration and experimentation, and (3) knowledge of science. Comparative studies of the impact of science and non-science television have demonstrated that viewers of science shows (e.g., *Cro* television program) were more likely to be interested in learning more science or doing science activities. Television with less of a narrative element, such as *Bill Nye the Science Guy* had less impact on viewers' attitudes. However, control group studies of the impact of *Bill Nye the Science Guy* found that viewers were significantly more capable of making observations and comparisons than non-viewers (Rockman et al., 1996). Thus, television can successfully model scientific processes for viewers. Some research results are not surprising, for example that older children tend to show greater gains in content knowledge. Other results, however, demonstrate the distinct impacts of this technology, such as effects that are strongest for girls, an audience typically underserved by technology.

Comparative study of the treatment of science across television genres (the narrative style of *Cro* versus the documentary style of *3-2-1 Contact*) found several significant differences (Fisch, 2004b). In the narrative format, scientific explanations were broken up and spread among multiple characters in contrast to the direct approach of the documentary format. Content was constrained in different ways. In the narrative format, content had to fit the setting (e.g., the *Ice Age*), but could be addressed on a scale that was impossible in the documentary format (e.g., *Giant Catapults*). The documentary format had the freedom to address a wider variety of content in more real-life contexts. A critical final distinction was the marketability of the narrative format television. These shows have been able to move beyond the more limited PBS audience to a larger and more diverse audience of Saturday morning cartoons, for example.

Although children benefit from viewing television by themselves, the benefits are generally said to improve if parents and children watch together (e.g., Reiser, Williamson, & Suzuki, 1988). These findings are similar to those based on family use of interactive exhibits in science museums. Using an exhibit or watching a television program alone provide accessible experiences to the child, but adult interaction can enrich or extend them (Fisch, 2004; Gleason & Schauble, 2000).

Finally, an enabling feature of the technology of television is its ease of use. Unlike the other technologies discussed in this chapter, children can watch television from an early age. Usability issues endemic to interactive websites, software, and exhibits do not apply to television technology in the same way (Fisch, 2004a).

CLASSROOM IMPLICATIONS

Research studies on technologies for informal learning can benefit classroom instruction in a variety of ways. After all,

> what's considered fun, good learning in informal settings could and should also be the norm in formal learning. Formal education could benefit from the emphases on personal involvement with ideas, open-style learning environments, and broad diversity among collaborating learners. (Strohecker & Butler, 2004, p. 151)

Work in informal domain can help demonstrate for formal educators the importance of a learner focus, motivation, and non-cognitive learning outcomes, as well as the value of inquiry-based and constructivist learning approaches. The "emergence of rich constructivist environments can be facilitated by the emergence of powerful technology tools" (Karagiorgi & Symeou, 2005, p. 24), which are more likely to be applied first in informal learning environments.

"Technology transfer" from informal to formal settings may involve modification of the technology or conversion into suitable formats. For example, StarLab (www.starlab.com), a small-scale inflatable version of a planetarium, can be used to support the Planetarium Activities for Student Success (PASS) program developed by the Lawrence Hall of Science (www.lhs.berkeley.edu/PASS/) and other activities tied to the curriculum. Web-based informal programs, such as citizen science, where volunteers carry out research-related tasks, may be transported, as in the case of Cornell Laboratory of Ornithology's *Classroom FeederWatch* (www.birds.cornell.edu/schoolyard/). Existing materials may be repurposed as enhancements to instruction. For example, teachers can record or purchase copies of *Cyberchase* television episodes on topics in math. The companion Web site offers teachers follow-up activities, interactive challenges, and video clips that connect the math topics to student interests and to National Council of Teachers of Mathematics (NCTM) standards. Similar educator support is provided for *Zoom, Dragonfly TV, NOVA*, and other educational television programs; many of these materials are accessible through PBS TeacherSource (www.pbs.org/teachersource/sci_tech.htm).

Science museums, schools and community-based organizations have employed various forms of technology to enhance after-school programs, which take advantage of a range of computer and web-based software applications (e.g., www.kineticcity.com), wireless and mobile communication devices, Geographic Information Systems (GIS), and Global Positioning Systems (GPS), for example. Development in this area has been stimulated by NSF's Information Technology Experiences for Students and Teachers (ITEST) program (see www2.edc.org/itestlrc for examples).

CHALLENGES TO THE FIELD

Because technology plays such a key role in most forms of informal learning, the question "how we *should* use technology—as opposed to how we *can* use it" (Parkes, 2004, p. 35, emphasis in the original) assumes great importance for science museums and the diverse panoply of technologies employed in other settings. Similarly, "the story of digital technologies in educational contexts has often been one of a solution in search of a problem" (Hawkey, 2004, p. 39). The limited but growing body of evidence from research and evaluation studies can help guide appropriate applications of technology based on the intended audience, content, and setting, as well as the design of the software or programming.

Educational research and evaluation involving technology must grapple with additional complexity beyond the general challenges faced by studies on informal learning mentioned earlier. The impact of the educational technology must be separated from other independent variables in settings that may not readily lend themselves to control groups or otherwise generalizable research. Furthermore, the interrelationship of hardware and software must be unraveled across a wide range of possibilities. The hardware may serve as the primary learning vehicle, as in a physically interactive exhibit, or it may be nearly invisible to the learner, as in a giant-screen theater. Similarly, software may be the primary driver of interaction, as on the Web, or it may be essentially absent, in the case of direct physical interaction. Also, the role played by the interface, the means by which visitors or users experience the technology, whether or not mediated by software, may be underestimated (Parkes, 2004). In each case, it is design that shapes the interaction of the technology with the user. That design may be guided by best practices (e.g., McLean & McEver, 2004) and by models of informal learning (e.g., Falk & Dierking, 2000). However, a much more comprehensive body of educational research is needed in general to provide an evidence-based foundation for the creative design and application of informal learning technologies. Furthermore, this body of knowledge will need to be translated into forms that can guide practitioners.

Further study is needed in many areas. For example, what is impact of the physical and social environment in mobile learning? How can the design of learning experiences best encourage social interactions that facilitate learning conversations? This issue is especially important for personalization technologies that may isolate the user, inhibiting social interaction, unless the experience is designed for collaboration or sharing. One possible direction is offered by findings from study of "smart toys," which show a potential for encouraging increased social interaction among children, along with providing rich learning interactions. (Plowman & Luckin, 2004) Another area that warrants further study is the role of gender and cultural differences in the use and impact of various informal learning technologies. Initial work indicates that girls and boys tend to prefer different types of online activities (Schaller, Allison-Bunnell, & Borun, 2005).

As the work cited here demonstrates, technology is no panacea for informal STEM education, and like all tools, has significant limitations. When the latest technology is employed for its ability to attract users, the technology itself may overshadow the educational application. Certainly users may still learn something about the technology itself, which can be of value, but not necessarily the intended outcome. In addition, an excess of technological novelty may lead to "cognitive overload" as visitors attempt to make sense of each new device (Allen, 2004). Of course, the simple act of using the technology does not necessarily lead to learning, which requires "minds-on" as well as "hands-on" interaction (Gregory,

1989). Especially if "high-tech," the technology may be complex to use and to understand. Its impact may be limited to "those already engaged in inquiry and who have sufficient prior experience to interpret the results of the technology" (Ansbacher, 1997, p. 3). From a practical stand-point, it is also challenging for institutions to introduce new technologies continually as a means for learner engagement as the pace of development and application to consumer products increases. If cutting-edge, or worse, "bleeding-edge," the technology itself may be prone to breakdown and require frequent maintenance, confounding attempts to study visitor interaction and behavior.

OPPORTUNITIES

Despite these types of challenges, informal learning technologies offer great potential that further research will only help to develop. Technologies increasingly enable personalization and customization, such as through visitor RFIDs in science museums. Portable and wireless devices bring mobility and remove place-based limitations, such as through web browsing with cell phones or PDAs, allowing "just in time" learning (Bransford, Brown, & Cocking, 2000). Similarly, the shift from broadcast to interactive media and "narrowcasting" makes possible highly targeted programming and information on demand at any time, such as via podcasts. These continuing trends will make possible greater connections across learning that occurs by different modes in different places at different times.

In response, place-based institutions must seek new strategic niches, which in turn can be facilitated by educational technologies. It has been suggested "the museum is the perfect place in which to produce a true physical hypermedia system" (Jones, 2002, p. 5). This direction is bolstered by the recommendation from a MacArthur Foundation-funded planning grant on digital learning and play for further research and development related to "new hybrid spaces that blend physical and virtual environments and digital tools," (Hsi, 2005, p. 4). These combinations of learning resources must support interactive social group learning experiences, as well as authentic experiences augmented by appropriate technologies. In this way, "the integration of real and virtual will provide further powerful learning opportunities" (Hawkey, 2004, p. 38), creating places for sharing personalized learning experiences.

Furthermore, emerging network-based technological applications also offer great potential for informal STEM learning. They will likely be influenced by the "nascent revolution" arising from the advanced networking and distributed scientific knowledge environments of the evolving cyberinfrastructure (NSF, 2003, p. 9). The NSF-supported Cyberinfrastructure for Education and Learning for the Future (CELF) initiative presents a vision of cyber-mediated learning that "will take place in the context of computationally augmented real-world environments, online communities of practice, interactive virtual environments, games, simulations, models, and audio/video/IM/SMS communications—not just in classrooms" (Computing Research Association, 2005, p. 9). The CELF vision also includes the intriguing notion of a lifelong learning chronicle (LLC) that would take advantage of digital technology to provide a qualitative and quantitative record of learning over time. In these ways, "informal learning may harness ubiquitous computing environments of the future by providing 'learning services' to people in formal, non-formal, and informal settings, and by helping people to manage their personal learning goals, projects, and informal learning activities" (Roberts et al., 2005, p. 10).

More educational research also will be needed to understand the impact of aggregation of learning experiences that combine various informal modes, whether they occur in hybrid forms at one time and place, such as augmented reality, or as asynchronous experiences across multiple platforms, such as the combination of *Cyberchase* TV shows and an associated web site or *ZOOM* TV programs and *ZOOMzone* museum exhibits. These latter approaches take a step towards creating learning communities that cut across the institutional borders of museums, libraries, television, and other informal learning resources, as well as the formation of an "ecology of learning" in which various types of organizations serve complementary niches. Studying the interaction of the learner with the associated informal technologies in a particular environment, along with learner-to-learner interaction, adds additional layers of complexity, but will be essential for maximizing educational impact. The potential for embedded monitoring within interactive technology may be assist such studies, assuming that privacy issues can be addressed.

CONCLUSION

Learning technologies in informal settings can provide enrichment and extension experiences through such means as museum field trips, after-school programs, and home-based Web and television that are complementary to classroom instruction in schools, which are "just part of a broader learning ecosystem. In the digital age, learning can and must become a daylong and lifelong experience" (Resnick, 2002, p. 36). Enriched activities outside the classroom such as these have been associated with higher scientific reasoning ability for students (Gerber, Cavallo,

& Marek, 2001). Studies identify pre- and postvisit preparation, teacher professional development, and integration with classroom learning as factors that reinforce the learning that takes place in these different settings (Anderson, Lucas, Ginns, & Dierking, 2000). Nevertheless, "further study of young people's informal learning with technology and its relationship to their formal training is both a research need and a gap" (Fitzgerald, 2005, p. 4).

Such studies, along with the work cited in this chapter, can help achieve more seamless student learning, creating "classrooms without walls" that break down distinctions now separating the formal and informal realms. In addition, they support the development of a science of learning that brings together "understandings of both informal and formal learning environments, drawing on the best features of all known learning environments to build the schools of the future" (Sawyer, 2006, p. 568). Research on technology-based informal learning certainly will not provide all the answers. However, it can suggest fruitful directions through an emphasis on the learner focus, role of motivation and context, customized just-in-time learning, social interaction, and home- and community-based activities (Sawyer, 2006). In so doing, emphasis on learners and learning can help guide the transformation of our schools in ways that make them most effective for preparing workers and citizens needed for our nation to be competitive in the twenty-first century.

ACKNOWLEDGEMENT

Any opinions, findings, and conclusions or recommendations expressed in this chapter are those of the authors and do not necessarily reflect the views of the National Science Foundation.

REFERENCES

Allen, S. (2004). Designs for learning: Studying science museum exhibits that do more than entertain. *Science Education, 88* (Suppl. 1), S17-S33.

Allison-Bunnell, S., & Schaller, D. T. (2005). From the physical to the virtual: Bringing free-choice science education online. In L. Tan Wee Hin & R. Subramaniam (Eds.), *E-learning and virtual science centers* (pp. 163-189). Hershey, PA: Idea Group.

Anderson, D., Lucas, K. B., & Ginns, I. S. (2003). Theoretical perspectives on learning in an informal setting. *Journal of Research in Science Teaching, 40*(2), 177-199.

Anderson, D., Lucas, K. B., Ginns, I. S., & Dierking, L. D. (2000). Development of knowledge about electricity and magnetism during a visit to a science museum and related post-visit activities. *Science Education, 84*(5), 658-679.

Andresen, L., Boud, D., & Cohen, R. (2000). Experience-based learning. In G. Foley (Ed.), *Understanding adult education and training* (2nd ed., pp. 225-239). Sydney, Australia: Allen & Unwin.

Ansbacher, T. (1997). If technology is the answer, what was the question? Technology and experience-based learning. *Hand to Hand, 11*(3), 3, 6.

Association of Science-Technology Centers. (2005). *2005 ASTC Sourcebook of Statistics and Analysis.* Washington, DC: Author.

Bandelli, A. (2005, June 15-18). Nest box cams: Site Critique. In *Web designs for interactive learning conference proceedings.* Retrieved July 4, 2007, from Web Designs for Interactive Learning Web site: http://www.wdil.org/conference/pdf/crt12z.pdf

Barstow, D. (2005, June 15-18). Great backyard bird count: Site Critique. In *Web designs for interactive learning conference proceedings.* Retrieved July 4, 2007, from Web Designs for Interactive Learning Web site: http://www.wdil.org/conference/pdf/crt12z.pdf

Bitgood, S. (2002). Environmental psychology in museums, zoos, and other exhibition centers. In R. Bechtel & A. Churchman (Eds.), *Handbook of environmental psychology* (pp. 461-480). New York: Wiley.

Bonney, R. (2005, June 15-18). Great backyard bird count: Site presentation. In *Web designs for interactive learning conference proceedings.* Retrieved July 4, 2007, from Web Designs for Interactive Learning Web site: http://www.wdil.org/conference/pdf/crt12z.pdf

Bransford, J., Vye, N., Stevens, R., Kuhl, P., Schwartz, D., Bell, P. et al. (2006). (2006). Learning theories and education: Toward a decade of synergy. In P. A. Alexander & P. Winn (Eds.), *Handbook of Educational Psychology* (2nd ed. pp. 209-244). Mahwah, NJ: Erlbaum.

Bransford, J. D., Brown, A. L., & Cocking, R. R. (Eds.). (2000). *How people learn: Brain, mind, experience, and school.* Washington, DC: National Academies Press.

Bruce, B. C., & Levin, J. A. (1997). Educational technology: Media for inquiry, communication, construction, and expression. *Journal of Educational Computing Research, 17*(1), 79-102. Retrieved July 5, 2007, from University of Illinois, Bertram Bruce Web site: http://www.isrl.uiuc.edu/~chip/pubs/taxonomy/taxonomy.pdf

Carlson, S., & Maxa, S. (1997). *Science guidelines for nonformal education. Cooperative Extension Service, Children, Youth and Family Network, CREES-USDA.* Retrieved July 4, 2007 from, http://www.cyfernet.org/science/4h590.html

Chapman, J. D., & Aspin, D. N. (2000) Lifelong learning: Concepts and conceptions. *International Journal of Lifelong Education, 19*(1), 2-19.

Colson, C. (2005). Accessing the microscopic world. *PLoSBiology, 3*(1), 27-29.

Computing Research Association. (2005). *Cyberinfrastructure for education and learning for the future: A vision and research agenda.* Retrieved July 4, 2007, from Computing Research Association Web site: http://www.cra.org/reports/cyberinfrastructure.pdf

Crowley, K., & Jacobs, M. (2002). Building islands of expertise in everyday family activity. In G. Leinhardt, K. Crowley, & K. Knutsen (Eds.). *Learning conversations in museums* (pp. 333-356). Mahwah, NJ: Erlbaum.

de Strulle, A. (2004). Differentiation of the causal characteristics and influences of virtual reality and the effects on learning at a science exhibit. *Dissertation Abstracts International, 65*(05A), 1745. (UMI Dissertation Information Service, No. 072699)

Dierking, L. D., Ellenbogen, K. M., & Falk, J. (2004). In principle, in practice: Perspectives on a decade of museum learning research (1994-2004). *Science Education, 88*(Supp. 1), S1-3.

Eberbach, C. (2006). *The Tech Tag Study* (Internal report). San Jose, CA: The Tech Museum of Innovation.

Eberbach, C., & Crowley, K. (2005). From living to virtual: Learning from museum objects. *Curator, 48*(3), 317-338.

Educational Technology Research & Assessment Cooperative. (2002). *Learning sciences and technology taxonomy* (Ver. 2). Retrieved July 4, 2007 from LESTER, Learning Science & Technology Repository: http://antioch.rice.edu/etrac/lester/

Ellenbogen, K. M. (Ed.). (2003). Sociocultural perspectives on museums. *Journal of Museum Education, 28*(1), 2.

Fallows, D. (2006). Browsing the Web for fun. *Pew Internet Project Data Memo.* Retrieved July 4, 2007, from Pew Internet &American Life Project Web site: http://www.pewinternet.org/pdfs/PIP_Surfforfun_Feb06.pdf

Falk, J. H., & Dierking, L. D. (2000). *Learning from museums: Visitor experiences and the making of meaning.* Walnut Creek, CA: AltaMira Press.

Falk, J. H., & Dierking, L. D. (2002). *Lessons without limit: How free-choice learning is transforming education.* Walnut Creek, CA: AltaMira Press.

Falk, J. H., Dierking, L. D., & Storksdieck, M. (2005, March). *A review of research on lifelong science learning.* Paper presented at The Status of Research on Learning Science Within Informal Education Settings, The National Academies, Washington, DC.

Falk, J., & Storksdieck, M. (2005). Using the contextual model of learning to understand visitor learning from a science center exhibition. *Science Education, 89,* 744-778.

Fisch, S. M. (2004a). *Children's learning from educational television: Sesame Street and beyond.* Mahwah, NJ: Erlbaum.

Fisch, S. M. (2004b). Characteristics of effective materials for informal education: A cross media comparison of television, magazines, and interactive media. In M. Rabinowitz, F. C. Blumberg, & H. T. Everson (Eds.), *The design of instruction and evaluation: Affordances of using media and technology* (pp. 3-18). Mahwah, NJ: Erlbaum.

Fisher, M. (1997). The effect of humor on learning in a planetarium. *Science Education (Informal Science Education—Special Issue), 81*(6), 703-713.

Fitzgerald, R. (2005). Understanding informal learning with technology: Insights for ICT integration. In P. Kommers & G. Richards (Eds.), *World conference on educational multimedia, hypermedia and telecommunication.* Retrieved July 4,

2007, from University of Canberra, Robert Fitzgerald-Digital Ecologies Web site: http://creative.canberra.edu.au/digital/docs/edmedia2005.pdf

Flagg, B. (2005). Beyond entertainment: Educational impact of films and companion materials. *The Big Frame, (Spring)*, 51-56.

Flagg, B. N. (1994). Visitors in front of the small screen. *ASTC News, 2*, 21-24.

Fleck, M., Frid, M., Kindberg, T, O'Brien-Strain, E., Rajani, R., & Spasojevic, M. (2002a). *Rememberer: A tool for capturing museum visits.* Retrieved July 4, 2007, from HP Web site: http://www.hpl.hp.com/techreports/2002/HPL-2002-187.html

Fleck, M., Frid, M., Kindberg, T., Rajani, R., O'Brien-Strain, E., & Spasojevic, M. (2002b). *From informing to remembering: Deploying a ubiquitous system in an interactive science museum.* Retrieved July 4, 2007, from HP Web site: http://www.hpl.hp.com/research/papers/2002/ubiquitous.html

Gerber, B. L., Cavallo, A. M. L., & Marek, E. A. (2001). Relationships among informal learning environments, teaching procedures, and scientific reasoning ability. *International Journal of Science Education, 23*(5), 535-549.

Gleason, M. E., & Schauble, L. (2000). Parents' assistance of their children's scientific reasoning. *Cognition and Instruction, 17*(4), 343-378.

Grandy, R. E., & Duschl, R. A. (2005). *Reconsidering the character and role of inquiry in school science: Analysis of a conference.* Presented at International History and Philosophy of Science and Science Teaching Group meeting in Leeds, England. Retrieved July 4, 2007, from Rice University Web site: http://www.ruf.rice.edu/~rgrandy/LeedsREGE.pdf

Gray, D. E. (1999). The internet in lifelong learning: Liberation or alienation? *International Journal of Lifelong Learning, 18*(2), 119-126.

Gregory, R. L. (1989). *The Nuffield Foundation Interactive Science and Technology Project.* London: Nuffield Foundation.

Haley Goldman, K., & Schaller, D. T. (2004, April). *Exploring motivational factors and visitor satisfaction in on-line museum visits.* Presented at Museums and the Web 2004, Arlington, VA.

Hawkey, R. (2004). Report 9: Learning with digital technologies in museums, science centres, and galleries. *NESTA Futurelab.* Retrieved July 4, 2007 from http://www.nestafuturelab.org/research/reviews/09_01.htm

Heath, C., & vom Lehn, D. (2003). Misconstruing Interaction. In M. Hinton (Ed.), *Interactive learning in museums of art and design.* London: Victoria and Albert Museum.

Heath, C., vom Lehn, D., & Osborne, J. (2005). Interaction and interactives: Collaboration and participation with computer-based exhibits. *Public Understanding of Science, 14*(1), 91-101.

Hein, G. E. (1995). The constructivist museum. *Journal of Education in Museums, 16*, 21-23.

Hsi, S. (2003). A study of user experiences mediated by nomadic web content in a museum. *Journal of Computer Assisted Learning, 19*, 308-319.

Hsi, S. (2005). *Digital-mediated experiences of next generation youth: Recommendations for future investment.* Retrieved July 4, 2007, from Exploratorium Web site: http://www.exploratorium.edu/research/digitalkids/OverviewHsi_DigitalKids.pdf

Hsi, S., & Fait, H. (2005). RFID enhances visitors' museum experiences at the Exploratorium. *Communications of the ACM, 48*(9), 60-65.

Jackson, P. W. (1968). *Life in classrooms.* New York: Holt, Rinehart & Winston.

Jensen, J. (1999, August). *Goals, server logs, and other arcane lore: An evaluation of two Canadian Web sites.* Proceedings of the Visitor Studies Association Conference, Chicago.

Jones, S. (2002). *The hypermuseum.* Paper presented at the 3rd Science Centre World Congress, Canberra, Australia. Retrieved July 4, 2007, from Online Culture Channel, Stephen Jones Web site: http://www.culture.com.au/brain_proj/

Kafai, Y. B., Fishman, B. J, Bruckman, A. S., & Rockman, S. (2002). Models of educational computing @ home: New frontiers for research on technology in learning. *Educational Technology Review, [Online serial], 10*(2), 52-68. Retrieved July 4, 2007 from, https://www.aace.org/pubs/etr/issue3/Kafai.pdf

Karagiorgi, Y., & Symeou, L. (2005). Translating constructivism into instructional design: Potential and limitations. *Educational Technology & Society, 8*(1), 17-27.

Kolb, D. A., Boyatzis, R. E., & Mainemelis, C. (2000). Experiential learning theory: Previous research and new directions. In R. J. Sternberg & L. F. Zhang (Eds.), *Perspectives on cognitive, learning, and thinking styles.* Mahwah, NJ: Erlbaum.

Leinhardt, G., & Knutson, K. (2004). *Listening in on museum conversations.* Walnut Creek, CA; Alta Mira Press.

Lewenstein, B. (2001). Who produces science information for the public? In J. H. Falk, (Ed.), *Free-choice science education: How we learn science outside of school* (pp. 21-43). New York: Teachers College Press of Columbia University.

Linn, M. C., Davis, E. A., & Bell. P. (2004). Inquiry and technology. In M. C. Linn, E. A. Davis, & P. Bell (Eds.) *Internet environments for science education* (pp. 3-28). Mahwah, NJ: Erlbaum.

Marty, P. F. (2005). 21st century museum visitors and digital museum resources. *Digital Library of Information Sciences and Technology.* Retrieved July 4, 2007, from Digital Library of Information Science and Technology Web site: http://dlist.sir.arizona.edu/1042/

McLean, K., & McEver, C. (Eds.). (2004). *Are we there yet: Conversations about best practices in science exhibition development.* San Francisco: Exploratorium.

Meredith, J. E., Fortner, R. W., & Mullins, G. W. (1997). Model of affective learning for nonformal science education facilities. *Journal of Research in Science Teaching, 34*(8), 805-818.

Moussouri, T. (1997). *Family agendas and family learning in hands-on museums.* Unpublished dissertation, University of Leicester, England.

Naismith, L., Lonsdale, P., Vavoula, G., & Sharples, M. (2005). Report 11: Literature review in mobile technologies and learning. *NESTA Futurelab.* Retrieved July 4, 2007, from http://www.nestafuturelab.org/download/pdfs/research/lit_reviews/futurelab_review_11.pdf

National Science Board. (2006). Science and technology: Public attitudes and understanding (Chapter 7). In *Science and engineering indicators 2006. Two volumes* (Vol. 1, NSB 06-01; Vol. 2, NSB 06-01A). Arlington, VA: Author. Retrieved July 4, 2007, from http://www.nsf.gov/statistics/seind06/

National Science Foundation. (2003). *Revolutionizing science and engineering through cyberinfrastructure: Report of the National Science Foundation Blue-Ribbon Advisory Panel on Cyberinfrastructure. CISE-051203.* Retrieved July 4, 2007, from http://www.nsf.gov/od/oci/reports/toc.jsp

National Science Foundation. (2006). *Informal Science Education (ISE) Program Solicitation* (NSF 06-520). Retrieved July 4, 2007, from http://www.nsf.gov/pubs/2006/nsf06520/nsf06520.htm

Newsom, C. V. (Ed.). (1952). *A television policy for education.* Washington, DC: American Council on Education.

Olson, S., & Loucks-Horsley, S. (Eds.) (2000). *Inquiry and the national science education standards: A guide for teaching and learning.* Washington, DC: National Research Council. Retrieved July 4, 2007, from http://books.nap.edu/catalog/9596.html

Opinion Dynamics Corporation. (2005). *More than one-third of Americans saw a giant screen film in 2004* (Press release). Retrieved July 4, 2007, from http://www.opiniondynamics.com/docs/Giant_Screen_Filmgoers_2004.pdf

Oppenheimer, F. (1975). The Exploratorium and other ways of teaching physics. *Physics. Today, B*(9), 9- 13.

Parkes, A. (2004). Employing new technology: Think about the interface. In K. McLean & C. McEver (Eds.), *Are we there yet: Conversations about best practices in science exhibition development* (pp. 35-37). San Francisco: Exploratorium.

Picard, R. W., Papert, S., Bender, W., Blumberg, B., Breazeal, C., Cavallo, D. et al. (2004). Affective learning: A manifesto. *BT Technology Journal, 22*(4), 253-269. Retrieved July 4, 2007, from Massachusetts Institute of Technology Web site: http://www.media.mit.edu/publications/<wbr>bttj/Paper26Pages253-269.pdf

Plowman, L., & Luckin, R. (2004). Interactivity, interface, and smart toys. *IEEE Computer, 37*(2), 98-100.

Reiser, R. A., Williamson, N., & Suzuki, K. (1988) Using "Sesame Street" to facilitate children's recognition of letters and numbers. *Educational Communication and Technology Journal, 36(*1), 15-21.

Rennie, L. J., Feher, E., Dierking, L. D., & Falk, J. H. (2003). Toward an agenda for advancing research on science learning in out-of-school settings. *Journal of Research in Science Teaching, 40*(2), 112-120.

Resnick, M. (2002). Rethinking learning in the digital age. In G. Kirkman, P. K. Cornelius, J. D. Sachs, & K. Schwab (Eds.), *The global information technology report: Readiness for the networked world* (pp. 32-37). New York: Oxford University Press. Retrieved October 28, 2007, from Massachusetts Institute of Technology, Media Lab Web site: http://llk.media.mit.edu/papers/mres-wef.pdf

Roberts, G., Aalderink, W., Cook, J., Feijen, M., Harvey, J., Lee, S., et al. (2005). *Reflective learning, future thinking: Digital repositories, e-Portfolios, informal learning, and ubiquitous computing* (SURF_ALT_ILTA White Paper). Retrieved July 4, 2007 from Association of Learning and Technology Web site: http://www.alt.ac.uk/docs/ALT_SURF_ILTA_white_paper_2005.pdf

Rockman *et al.* (1996). *Evaluation of the Bill Nye the Science Guy Television Series and Outreach.* Retrieved July 4, 2007 from Rockman *et al* Web site: http://www.rockman.com/projects/124.kcts.billNye/BN96.pdf

Rosenfeld, S. B. (1980). Informal learning in zoos: Naturalistic studies of family groups. *Dissertation Abstracts International, 41*(07). (University Microfilms No. AAT80-29566).

Roussou, M. (2000, March). Immersive interactive virtual reality and informal education. In C. Stephanidis (Ed.), *Proceedings of i3 spring days workshop on User Interfaces for All: Interactive Learning Environments for Children,* Athens, Greece. Retrieved July 4, 2007, from makebelieve, Maria Roussou Web site: http://www.makebelieve.gr/mr/www/mr_publications.html

Roussou, M. (2004). Learning by doing and learning through play: An exploration of interactivity in virtual environments for children. *ACM Computers in Entertainment, 2*(1), 1-23.

Roussou, M., Johnson, A., Moher, T., Leigh, J., Vasilakis, C., & Barnes, C. (1999). Learning and building together in an immersive virtual world. *Presence 8,* 247-263.

Sandifer, C. (2003). Technological novelty and open-endedness: Two characteristics of interactive exhibits that contribute to the holding of visitor attention in a science museum. *Journal of Research in Science Teaching, 40*(2), 121–137.

Sawyer, R. K. (2006). Conclusion: Schools of the future. In R. K. Sawyer (Ed.), *The Cambridge handbook of the learning sciences* (pp. 567-580). Cambridge, England: Cambridge University Press.

Scanlon, E., Jones, A., & Waycott, J. (2005). Mobile technologies: Prospects for their use in informal science learnin (Portable learning: experiences with mobile devices). *Journal of Interactive Media in Education* (Special Issue), Retrieved July 4, 2007 from http://jime.open.ac.uk/2005/25/

Schaller, D. T., Allison-Bunnell, S., & Borun, M. (2005, April). *Learning styles and on-line interactives.* Presented at Museums and the Web 2005, Vancouver, Canada. Retrieved http://www.archimuse.com/mw2005/abstracts/prg_280000829.html

Semper, R., Wanner, N., Jackson, R., & Bazley, M. (2000, April). *Who's out there? A pilot user study of educational web resources by the Science Learning Network.* Paper presented at Museums & the Web 2000, Minneapolis, MN. Retrieved July 4, 2007 from http://www.archimuse.com/mw2000/papers/semper/semper.html

Serrell, B., & Raphling, B. (1992). Computers on the exhibit floor. *Curator, 35*(3), 181-189.

Sharples, M. (2003) Disruptive devices: Mobile technology for conversational learning. *International Journal of Continuing Engineering Education and Lifelong Learning, 12*(5/6), 504-520. Retrieved July 4, 2007 from University of Nottingham, Mike Sharples Web site: http://www.lsri.nottingham.ac.uk/msh/Papers/ijceell.pdf

Soloway, E., Grant, W., Tinker, R., Roschelle, J., Mills, M., Resnick, M. et al. (1999). Science in the palm of their hands. *Communications of the ACM, 42*(8), 21-26.

Sosniak, L. (2001). The 9% challenge: Educating in school and society. *Teachers College Record,* p. 103.

Spadaccini, J. (2006). *Museums and the Web 2.0. CHIN Roundtable: E-learning in museums.* Retrieved July 4, 2007, from Canadian Heritage Information

Network Web site: http://www.chin.gc.ca/English/Knowledge-Exchange/ panels-jim-spadaccini.php

Storksdieck, M. (2005). *Field trips in environmental education.* Berlin, Germany: Berliner Wissenschafts-Verlag.

Strohecker, C., & Butler, D. (2004). The informal informing the formal to form new models of learning. In *Proceedings of ACM IDC04: Interaction Design and Children* (pp. 151-152). Retrieved July 4, 2007, from Carol Strohecker Web site: http://www.carolstrohecker.info/PapersByYear/2004/InformalInforming.pdf

St. John, M., & Perry, D. (1996). *An invisible infrastructure: Institutions of informal science education* (Vol. I). Washington, DC: Association of Science-Technology Centers.

Sweeney, J. K., & Lynds, S. E. (2001). Reform and museums: Enhancing science education in formal and informal settings. In J. Rhoton & P. Bowers (Eds.), *Professional development leadership and the diverse learner* (pp. 125-131). Arlington, VA: National Science Teachers Association Press.

Ucko, D. A. (1983). Technology: Chance or choice? A museum exhibit on the impact of technology. *Science, Technology & Human Values, 8*(3), 47.

Ucko, D. A. (1998). *Science City at Union Station: A new model for recreational learning.* Retrieved July 4, 2007 from Museums+more Web site: http://www.museumsplusmore.com/pdf_files/SCarticle.pdf

Ucko, D. A., Schreiner, R. & Shakhashiri, B. Z. (1986). An exhibit on everyday chemistry: Communicating chemistry to the public. *Journal of Chemical Education, 63*,1081.

U.S. Department of Education. (1991). *America 2000: An education strategy, Sourcebook.* Washington DC: Author.

Vallance, E. (1995). The public curriculum of orderly images. *Educational Researcher, 24*(2), 4-13.

vom Lehn, D., Heath, C., & Hindmarsh, J. (2005). *Rethinking interactivity: Design for participation in museums and galleries.* Retrieved July 4, 2007 from Interaction Design Centre Web site: http://www.idc.ul.ie/museumworkshop/Papers/vom%20lehnetAl.pdf

Yu, K. C. (2005). Digital full domes: The future of virtual astronomy education. *The Planetarian, 34*(3), 6-11.

CHAPTER 10

THE IMPACT OF TECHNOLOGY ON SCIENCE PRESERVICE PREPARATION AND IN-SERVICE PROFESSIONAL DEVELOPMENT

Craig A. Wilson

The chapter focuses on a review of research conducted on the effects of technology in science teacher education that has been completed over the past 10 years. Seven of the studies that were reviewed focused on the effects of technology on preservice teachers and the other five studies investigated the effects of technology on in-service teachers. All studies included in the review had at least moderate internal and external validity and all yielded positive results. The chapter closes with a discussion of implications of technology use for teacher preparation programs by science teacher educators.

INTRODUCTION

The use and application of technology in K-12 teacher education has changed rapidly over the last few years. In 1997 the National Council for

The Impact of the Laboratory and Technology on Learning and Teaching Science K-16, pp. 267–287

the Accreditation of Teacher Education (NCATE) noted most teacher education programs treated technology as an add-on, rather than as a tool to be integrated across the curriculum. By 2000, NCATE had incorporated technology requirements into each of the six accreditation standards and, as a result, preservice teachers are now expected to know how to integrate technology into their classrooms and teacher education faculty are expected to model technology integration strategies in content and methods classes.

The national science standards also prescribe a curriculum in which technology is an essential component in both what is taught and how it is taught (Nies, 2001). For example, *The Benchmarks for Scientific Literacy* (American Association for the Advancement of Science, 1993) state that 9th-12th grade students should be able to, "Use computer spreadsheet, graphing, and database programs to assist in quantitative analysis" (p. 291). *The National Science Education Standards* (National Research Council, 1996) specify that teachers are to "provide the opportunity for students to use contemporary technology as they develop their scientific understanding" (p. 45).

Science education faculty across the country have reacted to the recent standards documents by increasing the level of technology integration in their teacher preparation programs. The following section includes a few examples of teacher education programs that have integrated technology into their courses.

WHAT IS TECHNOLOGY?

The term "technology" can have a variety of meanings. Therefore, it may be beneficial to define "technology" at the outset in order to establish an adequate context for the ensuing discussion. While some researchers choose to define technology in the very narrow sense as referring only to computers, others use the term in a much broader sense. For example, Moursund and Bielefeldt (1999) use the phrase "information technologies," which they divide into several categories, including "technology-assisted learning" and "technology as a tool."

"Technology-assisted instruction" includes computer assisted instruction, computer assisted research, and distance learning. An example of technology-assisted instruction can be seen at East Carolina University, where a "face-to face" elementary science methods course has been transformed into a distance-learning course. For part of the instruction, preservice teachers in the distance-learning course view the Annenberg Case Study Videos online and then utilize Blackboard for discussions. Students are required to pose questions and/or responses at least 4 times a week. For assessment, students go to the Prentice-Hall online textbook to take chapter-reading quizzes and they submit their other assignments via e-mail.

Another example of technology-assisted instruction comes from Lehigh University in Pennsylvania, where a professor is using analysis of inquiry learning and teaching to integrate technology into his elementary and secondary science methods courses. First, he asks his students to evaluate a Web-based inquiry (WBI) science lesson using a WBI instrument developed by Bodzin and Cates (2002). Participation in the evaluation process enhances the students' understanding of inquiry-oriented teaching methods and also their awareness of important characteristics of WBI. Following the evaluation process, students develop and implement their own WBI science lesson in order to apply what they have learned.

"Technology as a tool" involves word processors, graphics packages, scanners, digital cameras, presentation programs, databases, and spreadsheets. Oregon State University offers a graduate level program in which students learn to use technology as a tool for teaching science and math in Grades 3-12. Participants take a technology foundations course at the beginning of the program in which they learn to teach with technology, rather than just learning about technology. In one activity, students use the Internet to determine the geographical location of a mystery town and then they develop an interactive presentation using PowerPoint to explain how they used technology to identify the town. In another activity, students use a pH probe to distinguish differences among various household products and then they compare the effectiveness of using probes with other data gathering methods. Following the technology foundations course, students are expected to plan, implement, and reflect on lessons that integrate technology in the other courses in the program. In a microteaching course they are expected to integrate technology into at least one of their four lessons.

TECHNOLOGY INTEGRATION IN SCIENCE PRESERVICE TEACHER EDUCATION

The purpose of this section is to provide a review of the research studies in science preservice teacher education over the last 10 years. These studies were located by using a variety of data sources, including ERIC, bibliographies, reference lists, and conference proceedings. A total of 16 studies were identified, but only 7 are included in this review because of internal and external validity concerns with the other studies. Internal validity has to do with the interpretability of the results, based on adequate control of independent variables and having at least one comparison group. External validity is contingent on the generalizability of the results, based on the range of characteristics (i.e., gender makeup, ability levels, ethnicity) of the subject pool. The seven studies that were retained had at least moderate to high internal and external validity and they have been subdivided into three categories—technology integration in science methods courses, technology integration in science content courses, and technology integration in field-based experiences (see Table 10.1).

Technology Integration in Science Methods Courses

The following studies dealt with the effects of technology integration in science methods courses. In an ethnographic study, Davis and Falba (1999) attempted to determine factors that had a positive effect on learning about technology integration in a science methods course. The subjects were 101 elementary preservice education students who volunteered to be part of the study from one of four sections of the course that was taught at a large urban university.

The study involved a combination of interviews, observation, and document analysis. The students learned how to use a variety of technologies during the course, including spreadsheets to record and analyze data gathered during hands-on activities and digital cameras to present the results of their activities using PowerPoint presentations. Data from the three sources were triangulated and the author concluded that there were two major factors which contributed to learning about technology integration: (a) instructor modeling of computer integration activities; and (b) students receiving technical, curricular, material, and emotional support throughout the semester.

The purpose of the Cullin and Crawford (2003) study was to determine the effects of using dynamic systems modeling software on prospective science teachers' understanding of modeling systems and their intentions to use modeling systems in their own classrooms. The subjects were 14 preservice secondary science teachers who were enrolled in a science methods course. During the course the subjects participated in an instructional module that focused on how scientists use models. The course instructor utilized a dynamic systems software program called Model-It, in order to teach the subjects how to build, design, and test real-world problems via computer models.

The authors employed an ex post facto survey design in which data were gathered via questionnaires and interviews. The evidence indicated that there were two positive results. First, the subjects' understanding of modeling systems improved. Second, subjects' intentions to teach using models (as opposed to teaching about models) increased. The researchers concluded that the use of dynamic systems software could have a positive effect on prospective science teachers.

The internal validity of this study is moderate, due to the fact that there was no control group used as a comparison. However, the use of a pretest/posttest design did allow the researchers to compare subjects' responses at the end of the study to their responses at the beginning of the study, and this had a positive effect on the internal validity of the study.

In 1996, Daniel focused on the effects of an interactive multimedia program titled, Classroom With a View (CView) on preservice teachers' understanding of constructivist teaching strategies. The subjects consisted of 27 education majors who were enrolled in four different methods courses—middle school science, middle school math, secondary school science, and secondary school math.

All subjects worked with the CView program as part of their methods course. The CView program presents 10 math and science lessons designed to model constructivist teaching strategies. Each lesson includes embedded questions that encourage the user to reflect on the teacher's role and children's perspective for each lesson.

In this ethnographic study, extensive data were gathered over a 10-week period via interviews with students before and after using CView, observations of students using CView, students' writings while using CView, and questionnaires completed after using CView. The author triangulated the results from these data sources in order to analyze the length and depth of the interactions and concluded that the CView program helped the subjects develop a greater understanding of constructivist teaching strategies. Even though the author used triangulation to analyze the results in this study, it would have been more instructive to compare the effects of live simulations to the effects of using the videotaped simulations in the CView program.

Barnett, Keating, Harwood, and Saam (2002) investigated the effects of using The Inquiry Learning Forum (ILF) on preservice teachers' levels of interaction and belief systems as they learn to teach science using inquiry-oriented activities. The ILF is an Internet-based professional development system that includes eight 5-minute video vignettes of inquiry-oriented lessons taught by in-service teachers. It allows instructors to set up their own "inquiry circle," with discussion questions, external links, and lesson plans.

The subjects were 78 students in 3 elementary science methods courses at 3 different universities. After subjects viewed the video vignettes, they participated in asynchronous discussion forums that were linked to the videos. The discussion forums included the preservice teachers, instructors, and in-service teachers.

Using an ex post facto survey design, the researchers gathered data regarding the discussion points posted by the subjects via the discussion forums. Analysis of the types of discussion points and number of responses to the discussion points indicated that ILF was effective in creating meaningful discourse related to inquiry-oriented activities and preservice teachers' belief systems. Additionally, the majority of subjects stated the participation in the ILF helped them to develop a better understanding of inquiry-based teaching

The purpose of the Bodzin and Park (1998) study was to determine the effects of a Web-based forum on collaboration. The subjects for the study were 22 preservice secondary science teachers who were enrolled in a 5-week science methods course followed by a 10-week student teaching internship during what was described as the professional semester.

Subjects were introduced to a Web-based forum called SciTeach during the on-campus instruction at the beginning of the semester. SciTeach provides a user-friendly on-line environment in which preservice teachers can dialogue with their methods instructors and university supervisors. One important difference between a Web-based forum and a LISTSERV or bulletin board is that the Web-based forum files messages by topics. Subjects were asked to use the Web-based forum to post at least three messages per week throughout the semester.

In this ex post facto study, subjects were surveyed at the end of the semester, in order to determine the potential benefits of the Web-based forum. One of the reported benefits was that the subjects received socioemotional support via the messages that they sent to their methods instructor and university supervisor, thus reducing the isolation that student teachers often encounter. A second benefit reported by the subjects was that they were able to exchange information with their methods professor and university supervisor pertaining to pedagogical issues and curricular materials in their respective classrooms. The authors concluded that a Web-based forum can provide important support for student teachers.

Technology Integration in Science Content Courses

One study was located that dealt with the effects of technology integration in science content courses. In 1999 Otero, Johnson, and Goldberg explored the roles that computers can play in the development of physics knowledge. The subjects were junior and senior preservice elementary teachers enrolled in several sections of a physics course for elementary education majors titled, "Constructing Physics Understanding" (CPU). The authors do not specify how many sections or how many subjects were involved in the study.

The subjects worked in small groups to participate in activities that were presented to them on the computer screen. After the activities, subjects designed and implemented their own hands-on activities to test the predictions they had made, based on the results of the computerized activities. Finally, the subjects applied what the authors described as their "consensus ideas" by using them to solve problems.

This study used an ethnographic design to gather data through videotapes of the small group activities and large group discussions and interviews with subjects outside of class. The videotapes were transcribed and analyzed in a "variety of ways" (Otero et al., 1999, p. 61) and the authors came to the following conclusions regarding the role of the computer in concept formation: (a) the computerized physics activities supported collaboration as the subjects worked in their small groups, and (b) the computerized physics activities fostered the development of "explanatory models" (p. 87) as the subjects performed the computer activities and designed their own experiments. The internal validity of this study is only moderate, as it appears that the results were not triangulated. The author only states that they used a "variety of ways" to analyze the results.

Technology Integration in Field-Based Experiences

One study on the effects of technology in field-based experiences is presented here. Barclay-McLaughlin and Nonis (2001) conducted a study of the relationship between mentor teachers' and preservice teachers' use of technology during a 1-year internship. The subjects were 15 elementary education majors who spent one year in an urban (K-5) public school that emphasized science, math, and technology. The study utilized an ex post facto survey design that involved interviews with subjects at the end of the internship.

The researchers reached two primary conclusions, based on the results of interviews. First, mentor teachers used technology in two ways—as a whole class instruction tool and as a center for small group drill-and-practice. Second, they determined that there was a positive relationship between the mentor teachers' use of technology and the subjects' use of technology. Subjects who were placed in classrooms where mentor teachers used technology on a regular basis used technology more in their lessons than subjects who were placed with mentor teachers who used technology only occasionally.

TECHNOLOGY INTEGRATION IN
SCIENCE IN-SERVICE TEACHER EDUCATION

This section will focus on research involving in-service teachers in science teacher education during the last 10 years. As with the studies in preservice science teacher education, these studies were located by using a variety of data sources, including ERIC, bibliographies, reference lists, and conference proceedings. Six studies were found and five out of the six had at least moderate internal and external validity and could be included here (see Table 10.1). The five research studies have been subdivided into two categories—summer professional development programs and school year professional development programs.

Summer Professional Development Programs

There were three studies that investigated the effects of technology integration in summer professional development programs. Dori and Barnes (1994) investigated the effects of an in-service technology-training module on the attitudes of chemistry teachers. This study was included in the review, even though it exceeded the 10 year limit, because it was one of only two studies that utilized a quasi-experimental design. The subjects in this study were 66 participants of a summer training program in Israel.

The researchers utilized a quasi-experimental pretest/posttest control group design. The 39 subjects in the experimental group participated in a computer assisted instruction (CAI) module on polymers that provided opportunities for mastery learning and enrichment. The 27 subjects in the control group did not experience any topics related to CAI, although the authors do not describe the method of instruction for the control group. The subjects responded to an attitude inventory as a preassessment at the beginning of the study and as a postassessment at the end of the study.

The results of the study indicated that the experimental group (CAI) had more positive attitudes toward CAI and the use of computers than the control group. In a follow-up survey, subjects in the experimental group stated that they planned to use the polymer module in their own classrooms. They were especially focused on the models, animation, and visual effects presented.

Gammon, Hutchison, Waller, and Tolbert (1999) wanted to find out how a networking project affected teachers' understanding about how to use technology and the amount of time they devote to teaching science using the Internet and hands-on activities. The subjects in the study were 116 K-8 teachers who participated in a 2-week summer program that was designed to increase the use of technology in their science classrooms. There were two components in the project. First, the subjects learned how to communicate with each other with a computer network. Second, the subjects conducted chemistry and physical science activities at two different sites and then they shared results with their "electronic lab partners" at the other site, using digital photographs, graphs, and text. The teachers were asked to design Internet-based science experiments that they could conduct with their students during the coming school year.

The researchers employed a single-group design in which pre- and postcourse data were gathered. The long term results indicated that many of the subjects had become technology leaders in their schools and that they were also teaching a great deal more Internet-based and hands-on science in their classrooms then they were before the project. The authors concluded that the project was successful on several levels.

Rye (2001) investigated the effects of electronic concept map instruction on in-service teachers' concept mapping skill levels and their desire to use concept-mapping software in their own classrooms. The 18 subjects in this study were science teachers who had enrolled in a summer institute titled Health Sciences and Technology Academy (HSTA). The author defined concept mapping as "visual/spatial representations of ideas and concept interrelationships" (p. 224). One of the primary purposes of concept mapping was to help the learner relate new ideas to previously learned concepts. The subjects were taught how to use an electronic concept-mapping program and then partnered with university faculty to lead sessions in which public school students developed concept maps with the concept mapping program. The subjects also conducted in-service workshops with faculty in their own schools.

This was an ex post facto survey in which a wide range of data (electronic conferences, lesson plans, and institute evaluations) were gathered and then analyzed through a triangulation process. At the end of the institute, subjects reported that the project had increased their concept mapping skill levels and most of the subjects requested a site license so they could use the software when they returned to their classrooms.

School Year Professional Development Programs

Two studies investigated the effects of technology in school year professional development programs. Anfara, Danin, Melvin, and Dillner (2000) conducted a study to determine the effects of a technology-oriented professional development program on secondary science teachers' classroom practices and the science achievement of their students. The subjects consisted of 85 teachers and 294 students who were involved in a "Science Van Project" developed by the Delaware State Department of Education.

This project offered weekend science workshops and university coursework to the participating teachers. In addition, two science "specialists" visited secondary classrooms across the state, bringing laptop computers that interfaced with data-collecting probes. Each specialist spent between 4 and 8 days in the selected classrooms, working as a team with the classroom teacher to demonstrate the proper use of the technology and model inquiry-oriented methods.

In this ex post facto survey, the researchers used a variety of data sources to evaluate the Science Van project, including questionnaires, interviews, observations, and pre/posttests. The teachers involved in this study demonstrated a positive increase in their use of technology and inquiry-oriented methods and their students demonstrated positive gain scores on all four-unit tests. Thus, it was concluded that the Science Van Project had a positive effect on the classroom practices of the teachers and on the achievement of the students.

In another school year professional development study, Annetta and Shymansky (2006) wanted to determine the effects of three different distance learning strategies on the science achievement of elementary teachers. The three strategies were described as "live," "video," and "Web." A total of 94 teachers who were involved in a summer science workshop volunteered to participate in this study. The volunteers were stratified based on their scores on a science content test and then equal numbers of high, middle and low scorers were assigned to each of the three treatment groups.

All three groups attended sessions at remote sites, located in rural areas. The live group watched real-time audio/video broadcasts that were presented by a science expert and then they participated in a discussion with the presenter, followed by a discussion with a facilitator. The video group watched videotapes of those same broadcasts and discussions a week after they were originally aired, and then they also participated in a discussion with a facilitator. The Web group watched a streaming video of the presentations only, and they did not have a facilitator present for their discussion. All subjects were given a science content posttest and it was determined that the live strategy was more effective than the video strategy and the Web strategy and that the video strategy was more effective than the Web strategy.

This study utilized a quasi-experimental posttest only nonequivalent group design, and so there is reason to question internal validity, due to the possible confounding of the science content background of the subjects and the treatment. However, the fact that the researchers stratified the subjects, based on their science content scores compensates for this concern, as it could be argued that the groups were similar to each other, even though they were not randomly assigned. Another area of concern is that the live and video groups experienced the discussions with the presenter and they also had facilitators to keep them on task and continue the discussions, while the Web group did not get to experience the discussions with the presenter, nor did they have a facilitator with them to help keep them focused. Both of these factors may have had a negative impact on the results of the study, and so it is difficult to determine if it was the presentation mode or the use of a facilitator that caused the differences in the science content scores of the Web group.

SUMMARY OF RESULTS

A total of 12 research studies were presented in this chapter. Seven of the studies dealt with the effects of technology on preservice science teachers and the other five studies focused on the effects of technology on in-service science teachers. Table 10.1 provides an overview of the 12 studies.

As can be seen from the Table 10.1, technology had a positive impact in all 12 of the studies. Research findings in preservice science teaching indicated that the following independent variables had a positive effect: instructor modeling of computer integration, the use of modeling software, interactive video presentations of model lessons, Internet-based inquiry-oriented teaching vignettes, a Web-based forum, computerized physics activities, and mentor teacher use of technology. The following independent variables had a positive effect on in-service teachers: a computer-assisted instructional module, Internet-based science experiments, electronic concept mapping, a technology-oriented professional development program, and distance learning science presentations.

Three studies utilized an ethnographic design, six involved ex post facto surveys, and three were quasi-experiments. Internal validity was moderate in three of the studies and at least moderately strong in the other nine studies and external validity was at least moderately strong in all of the studies.

IMPLICATIONS

One of the implications that can be drawn from the information presented in this chapter is that teacher preparation institutions should find ways to increase the integration of technology in their science education

Table 10.1. Characteristics of the 12 Reviewed Studies Investigating the Impact of Technology on Science Preservice and In-Service Teacher Education Programs

Title/Author/Date	Subjects	Research Design	Major Findings	Internal/External Validity
Integrating Technology in Elementary Preservice Teacher Education: Orchestrating Scientific Inquiry in Meaningful Ways (Davis, 1999)	Preservice elementary teachers science methods course	Ethnographic	Two factors contributed to learning about technology integration: instructor modeling and students receiving support.	
Using Technology to Support Perspective Teachers in Learning and Teaching About Scientific Models (Cullin & Crawford, 2003)	Preservice secondary science teachers science methods course	Ex post facto survey	The use of modeling software to teach subjects how scientists use models improved their understanding of models and their intentions to use models in their classrooms.	Internal—moderate, due to lack of control group for comparison.
Helping Beginning Teachers Link Theory to Practice: An Interactive Multimedia Environment for Mathematics and Science Teacher Preparation (Daniel, 1996)	Preservice middle school and secondary science and math teachers science and math methods courses	Ethnographic	Interactive video presentations of model math and science lessons resulted in greater understanding of constructivist teaching strategies.	
Using Emerging Technologies to Help Bridge the Gap Between University Theory and Classroom Practice (Barnett et al., 2002)	Preservice elementary teachers science methods course	Ex post facto survey	Viewing Internet-based inquiry-oriented teaching vignettes resulted in meaningful discourse and enhanced subjects' understanding of inquiry-based teaching.	

Study	Population/Context	Methodology	Findings	Validity
A Study of Preservice Teachers' Interactions With a Web-based Forum (Bodzin & Park, 1998)	Preservice secondary science teachers science methods course and student teaching semester	Ex post facto survey	A Web-based forum provided emotional and instructional support for subjects.	Internal—moderate due to lack of triangulation of results
How Does the Computer Facilitate the Development of Physics Knowledge by Prospective Elementary Teachers? (Otero &Goldberg, 1999)	Preservice elementary teachers physics course	Ethnographic	Computerized physics activities supported collaboration and fostered the development of explanatory models.	
Preservice Teachers Experiences in a Technology-Rich Urban K-5 School Setting (Barclay-Mclaughlin & Nonis, 2001)	Preservice elementary teachers field-based experience	Ex post facto survey	There was a positive relationship between technology used by mentor teachers and technology used by subjects.	
In-service Chemistry Teachers Training: The Impact of Introducing Computer Technology on Teacher's Attitudes (Dori & Barnes, 1994)	In-service chemistry teachers summer professional development program	Quasi-experimental pretest/posttest control group design	A computer assisted instructional module had a positive effect on subjects' attitudes toward computer assisted instruction	
The Idaho K-8 Networking Project: Using the Internet to Improve K-8 Science Instruction (Gammon et al., 1999)	In-service K-8 teachers summer professional development program	Single group pretest/posttest	Subjects who learned to design Internet-based science experiments became technology leaders in their schools and taught more Internet-based science activities.	

Table continues on next page.

Table 10.1. Continued

Title/Author/Date	Subjects	Research Design	Major Findings	Internal/ External Validity
Enhancing Teachers' Use of Technology Through Professional Development on Electronic Concept Mapping (Rye, 2001)	In-service science teachers summer professional development program	Ex post facto survey	Learning to use an electronic concept mapping program enhanced the subjects' concept mapping skill levels and their intent to use the program in their own classrooms.	
Traveling Road or Professional Development? A Professional Development Science Project on Wheels (Angara et al., 2000)	In-service secondary science teachers school year professional development program	Ex post facto survey	A technology-oriented professional development program had a positive effect on subjects' use of technology and inquiry-oriented methods and their students' science achievement	
Investigating science learning for rural elementary school teachers in a professional development project through distance-education strategies (Annetta & Shymansky, 2006)	In-service elementary teachers school year professional development program	Quasi-experimental posttest only nonequivalent control group	Distance learning via live broadcasts and discussions had a more positive effect on subjects' science achievement than watching video-taped and video-streaming presentations.	Internal—moderate due to the possible confounding of two independent variables.

programs. A second implication is that teacher educators should conduct and publish research on the effectiveness of technology integration. Each of those implications will be discussed below.

Increase the Integration of Technology in Science Education Programs

One suggested way to increase the use integration of technology is to develop and fund a technology plan. According to the National Survey of Information Technology in Teacher Education, the majority of institutions do not have a plan for how they intend to integrate technology in their teacher education program (Moursund & Bielefieldt, 1999). In order for technology integration to occur, it is essential that institutions develop an effective technology plan.

The plan should involve the formation of a technology committee, composed of science education faculty and administrators. The technology committee should follow a bottom-up rather than a top-down model because the most effective technology-integration programs rely heavily on collaboration among the constituents, in order to foster a spirit of ownership. With ownership comes a desire to take a chance on innovative programs, as ideas percolate up from the technology committee. Without ownership, the innovative plans quickly go by the wayside, because faculty may view the innovations as an unnecessary addition to their workloads.

Furthermore, the technology plan should be funded, because funding provides an impetus for science education faculty to bring about major and sustained changes. If funding is not available, technology committees at the school and department level have little to offer to develop or sustain any meaningful change. Funding may be used to provide release time for faculty to develop technology integration activities and to support faculty who wish to attend technology-oriented professional conferences. Funding may also be used to bring technology consultants to the college and university campus to work with science education faculty and to purchase up-to-date technology resources.

Another suggested way to increase the integration of technology is to incorporate guidelines for technology use. The technology plan adopted by the institution should help science education faculty learn how to effectively model technology integration in their courses by incorporating prescribed guidelines. Representatives from National Association for research in Science Teaching, National Science Teachers Association, and Association of Science Teacher Educators met at the National Technology Leadership Retreat in order to develop guidelines for technology use (Lederman & Niess, 2000, pp. 2-3). The five guidelines that were developed during the retreat are paraphrased below. Also, included with the guidelines are examples of how technology might be incorporated into a science education program (Henriques, 2002).

1. Science teachers should be given opportunities to use new technology skills as they are learned, rather than learning those skills in isolation. For example, presentation software can be used to introduce an inquiry-oriented activity with photographs of the set-up.

2. Science teachers should be taught how to use technology to get students actively involved as they learn about essential concepts, rather than as a tool for memorizing facts. An example would be using spreadsheets to process and analyze data that have been gathered through an experiment.

3. Science teachers should learn how to use technology to go beyond what could be taught without technology rather than simply using technology to complete activities that could be done just as well without technology. For example, probeware can be set to collect temperature data over a long period of time or when no one is available to collect the data.

4. Technology should be used to make abstract science concepts more meaningful through visualization. For example, simulation programs can be used to make complex topics such as moon phases more understandable to students.

5. Science education courses should help teachers understand that technology is a tool for learning about science and also a tool for solving societal problems. For example, water quality readings could be taken by students living at various places along the length of the Mississippi River and then shared via the Internet in order to investigate potential sources of pollution.

Conduct and Publish Research on the Effectiveness of Technology Integration

Additional research is needed in order to determine the impact of technology on preservice preparation and in-service professional development, due to the small number of studies that have been completed over the last 10 years. After conducting an extensive search of the literature, only 12 studies were located for review in this chapter.

Another reason additional research is needed is due to the fact that few experimental studies have been conducted on this topic over the last 10 years. Most of the studies have been either ethnographic or ex post facto surveys. Therefore, future research should involve the manipulation of an experimental variable and use at least one control group for a comparison, in order to add to the experimental data base.

CONCLUSION

Today's K-12 students face a future in which technology will have an ever-increasing role and so it is essential that teachers prepare technologically literate students. However, according to Bell and Bell (2002) "most experienced teachers do not feel comfortable using technology in their instruction" (p. 2). According to NCATE (2000), only about 20% of teachers report that they feel prepared to integrate technology.

Teacher education programs are often blamed for this lack of preparation in technology. For example, Faison (1996) states, "many practicing teachers feel that they have not had adequate training to help them use technology effectively" (p. 57). A U.S. Congress Report (1995) states that in most institutions, "technology is not central to the teacher preparation experience" (p. 165). Bell and Bell (2002) note that most teachers have "never had coursework modeling the pedagogy of using technology in the context of science teaching" (p. 2).

Clearly, there is a great need for science teacher preparation programs to begin producing teachers who are technologically literate (Anderson & Borthwick, 2002). This will require a major shift from merely teaching about technology in stand alone courses to also teaching with technology in science courses. Moursund and Bielefeldt (1999) point out that stand alone courses may be necessary in certain contexts, such as bringing students up to speed on basic skills; however, they also state that teachers who are trained in programs which use "technology in all courses would be better prepared to use instructional technology" (p. 24). Settlage, Odom, and Pederson (2004) conclude, "If technology is integrated into a science methods course, then it seems more likely that the next generation of teachers will incorporate technology into their science teaching" (p. 3).

The solution seems very straightforward—science teacher education faculty should model a variety of technology tools in their content and methods courses. However, Colburn (2000) points out a very important principle when he states, "Before we—the professorate—can teach prospective K-12 teachers how to best use today's computer technology, we must be skilled *ourselves* at its use" (p. 2). Settlage, Odom, and Pedersen (2004) conducted a study of members of the Association of Science Teacher Educators (ASTE) and found that science teacher education faculty believes they still have a lot to learn about the uses of technology. This is why Colburn (2000) states that it will take massive professional development over the next few years in order for more faculty to share the technology education burden

The need for professional development is heightened as we consider the rapid and dramatic changes in technology. According to Flick and Bell (2000), the growth of digital technology has created a revolution in science education that is similar to the hands-on movement of the 1960s and is more pervasive than any previous curricular innovation in science. We have only begun to understand the tremendous potential of the technology that is currently available and technology is changing so rapidly that even science educators who are already technologically literate will need to be part of the concerted effort to increase their technology literacy.

This chapter closes with a call for institutions to dramatically increase professional development opportunities for their science education faculty who need to learn how to integrate technology into their classrooms. The professional development program should provide the time, equipment, and training they need for technology integration, as these were identified by teacher education faculty as the three greatest barriers to technology integration (Beggs, 2000). Science education faculty should then conduct and report on scientific research regarding the impact of technology on preservice and in-service teachers. This will allow us to plan and implement research-based practices in our science content and methods classrooms and thus take science education to levels that we would have never thought possible.

REFERENCES

American Association for the Advancement of Science. (1993). *Benchmarks for scientific literacy.* New York: Oxford University Press.

Anderson, C. L., & Borthwick, A. (2002). *Results of separate and integrated technology instruction in preservice training.* East Lansing, MI: ERIC Document Reproduction Service No. ED475921.

Anfara, V. A., Jr., Danin, S. T., Melvin, K., & Dillner, H. (2000). *Traveling road show or professional development? A professional development science project on wheels.* Paper presented at the annual meeting of the American Education Research Association in New Orleans, LA.

Annetta, L. A., & Shymansky, J. A. (2006). Investigating science learning for rural elementary school teachers in a professional-development project through distance-education strategies. *Journal of Research in Science Teaching 43*(10), 1019-1039.

Barclay-McLaughlin, G., & Nonis, A. (2001). Preservice teacher's experiences in a technology-rich urban k-5 school setting. *Society for Information Technology International Conference, 2001*(1), 1545-1546.

Barnett, M., Keating, T., Harwood, W., & Saam, J. (2002). Using emerging technologies to help bridge the gap between university theory and classroom practice: Challenges and successes. *School Science and Mathematics, 102*(6), 229-313.

Beggs, T. A. (2000). *Influences and barriers to the adoption of instructional technology.* Paper presented at the Mid-South Instructional Technology Conference in Mufreeboro, TN.

Bell, L., & Bell, R. (2002). Invigorating science teaching with a high-tech, low cost tool. *Edutopia.* Available at http://www.glef.org

Bodzin, A., & Cates, W. (2002). *Web-based inquiry for learning science: Instrument manual* (Ver. 1.0). Available at /techtoolarticle.html April 14, 2005, from http://www.lehigh.edu/~amb4/wbi/beta/beta2.pdf

Bodzin, A., & Park, J. (1998). A study of preservice science teachers' interactions with a Web-based forum. *Electronic Journal of Science Education, 3*(1). [Online].

Colburn, A. (2000). Changing faculty teaching techniques: A response to Flick and Bell. *Contemporary Issues in Technology and Teacher Education, 1*(1), 64-65.

Cullin, M., & Crawford, B. A. (2003). Using technology to support prospective science teachers in learning and teaching about scientific models. *Contemporary Issues in Technology and Teacher Education, 2*(4), 409-426.

Daniel, P. (1996). Helping beginning teachers link theory to practice: An interactive multimedia environment for mathematics and science teacher preparation. *Journal of Teacher Education 47*(3), 197-204.

Davis, K. S., & Falba, C. (1999). *Integrating technology in elementary preservice teacher education: Orchestrating scientific inquiry in meaningful ways.* Paper presented at the annual meeting of the American Education Research Association. Montreal, QB.

Dori, Y. J., & Barnes, N. (1994). *In-service chemistry teachers training: The impact of introducing computer technology on teachers' attitudes.* Paper presented at the annual meeting of the National Association for Research in Science Teaching, Anaheim, CA.

Faison, C. C. (1996). Modeling instructional technology use in teacher preparation: Why we can't wait. *Educational Technology, 36*(5), 57-59.

Flick, L., & Bell, R. (2000). Preparing tomorrow's science teachers to use technology: Guidelines for science educators. *Contemporary Issues in Technology and Teacher Education, 1*(1), 39-60.

Gammon, S. D., Hutchinson, S. G., Waller, B. E., & Tolbert, R. W. (1999). The Idaho K-8 Networking Project: Using the Internet to improve K-8 science instruction. *Journal of Chemical Education, 76*(5), 708-713.

Henriques, L. (2002). Preparing tomorrow's science teachers to use technology: An example from the field. *Contemporary Issues in Technology and Teacher Education, 2*(1), 13-18. [Online]

Lederman, N., & Niess, M. (2000). Technology for technology's sake or for the improvement of teaching and learning. *School Science and Mathematics, 100*(7),345-348.

Moursund, D., & Bielefieldt, T. (1999). *Will teachers be able to teach in a digital age? A national survey on information technology in teacher education.* Santa Monica, CA: Milken Exchange on Education Technology.

National Council for the Accreditation of Teacher Education. (1997). *Standards, Procedures and policies for the accreditation of professional education units.* Washington, DC: Author.

National Council for the Accreditation of Teacher Education. (2000). *Standards 2000.* Available at http://www.ncate.org/2000/2000stds.pdf

National Research Council. (1996). *National science education standards.* Washington, DC: National Academy Press.

Niess, M. L. (2001). A model for integrating technology in preservice science and mathematics content-specific teacher preparation. *School Science and Mathematics, 101*(2), 102-109.

Otero, V. K., Johnson, A., & Goldberg, F. (1999). How does the computer facilitate the development of physics knowledge by prospective elementary teachers? *Journal of Education, 181*(2), 57-89.

Rye, J. A. (2001). Enhancing teachers' use of technology through professional development on electronic concept mapping. *Journal of Science Education and Technology, 10*(3), 223-235.

Settlage, J., Odom, A. L., & Pedersen, J. E. (2004). Uses of technology by science education professors: Comparisons with teachers: Uses and the current versus desired technology knowledge gap. *Contemporary Issues in Technology and Teacher Education, 4*(3), 299-312.

U.S. Congress. (1995). *Teachers and technology: Making the connection.* Washington, DC: U.S. Government Printing Office.

BIBLIOGRAPHY

American Association of University Women. (2000). *Tech-savvy: Educating girls in the new computer age.* Washington, DC: Author.

Avraamidou, L., & Zembal-Saul, C. (2002). *Using a web task to make prospective elementary teachers' personal theorizing about science teaching explicit.* Paper presented at the Annual International Conference of the Association for the Education of Teachers of Science in Charlotte, NC.

Boone, W. J., & Gabel, D. (1994). Computers and preservice elementary science teacher Education. *Journal of Computers in Mathematics and Science Teaching, 13*(1), 17-42.

Cavanaugh, C. (2003). Information age teacher education: Educational collaboration to prepare teachers for today's students. *TecTrends, 47*(2), 24-27.

Ivers, K. S. (2002). *Changing teachers' perceptions and use of technology in the Classroom.* Paper presented at the annual meeting of the American Education Research Association. New Orleans, LA.

Knapp, L. R., & Glenn, A. D. (1996). *Restructuring schools with technology.* Boston: Allyn & Bacon.

Lehman, J. R. (1995). Microcomputers and the preparation of secondary science teachers: An eight year follow-up. *School Science and Mathematics, 95*(2), 69-77.

Linn, M. C. (2003). Technology and science education: Starting points, research programs, and trends. *International Journal of Science Education, 25*(6), 727-758.

Loschert, K. (2003). High-tech teaching. *Tomorrow's Teachers, 2003*(9), 2-5.

MacKenzie, J. (1999). How *teachers learn technology best*. Bellingham, WA: FNO Press.

McKay, M. & McGrath, B. (2000). Creating Internet-based projects: A model for Teacher professional development. *T.H.E. Journal, 27*(11), 114-124.

National Center for Educational Statistics. (1999). *Teacher quality: A report on teacher preparation and qualifications of public school teachers*. Washington, DC: Author.

Norby, R. N. (2002). *A study of changes in attitude towards science in a technology based K-8 preservice preparation science classroom*. Paper presented at the annual meeting of the National Association for Research in Science Teaching. New Orleans, LA.

Office of Technology Assessment. *Teachers and technology: Making the Connection* (OTA-HER-616). Washington, DC: U.S. Government Printing Office.

Schmidt, W. H., Raizen, S. A., Britton, E. D., Bianchi, L. J., & Wolfe, R. G. (1997). *Many visions, many aims: A cross-sectional investigation of curricular intentions in school science*. Dordrecht/Boston/London: Kluwer.

Snyder, L., Aho, A. V., Linn, M. C., Packer, A., Tucker, A., Ullman, J. & Van Dam, A. (1999). *Be FIT! Being fluent with information technology*. Washington, DC: National Academy Press.

Weinburgh, M., Smith, L., & Smith, K. (1997). Preparing preservice teachers to use technology in teaching math and science. *Tech Trends, 42*(5), 43-45.

Willis, J., Thompson, A., & Sadera, W. (1999). Research on technology and teacher education: Current status and future directions. *Educational Technology Research and Development, 47*(4), 29-45.

Yamagata-Lynch, L. C. (2003). How a technology professional development program fits into teacher's work life. *Teaching and Teacher Education 19*(6), 591-607.

ABOUT THE AUTHORS

Saouma BouJaoude was a science teacher for 13 years in Lebanon before he joined the University of Cincinnati where he completed a doctorate in Curriculum and Instruction with emphasis on science education in 1988. He taught in the United States for 5 years then returned to Lebanon in 1993 where he is now professor of science education and chair of the Department of Education at the American University of Beirut. Dr. BouJaoude's research interests include the nature of science, curriculum, and teaching methods. Reports of his research appeared in the *Journal of Research in Science Teaching, Science Education, International Journal of Science Education, School Science and Mathematics,* and *School Science Review* among other publications. He served as international coordinator of the National Association for Research in Science Teaching (NARST) and has consulted on projects in Lebanon, Egypt, Yemen, Oman, Saudi Arabia, Qatar, United Arab Emirates, and Jordan. He can be contacted at Department of Education, American University of Beirut, P.O. Box 11-0236, Bliss Street, Beirut, Lebanon or boujaoud@aub.edu.lb

Jale Cakiroglu is an associate professor of science education at Middle East Technical University, Turkey. She received her doctoral degree in curriculum and instruction from Indiana University with emphasis on science education. Since 2000, when she returned to Turkey, she has been teaching graduate and undergraduate courses in teacher training including methods of science teaching and instructional planning and evaluation, and curriculum in elementary science education. Her major research interests include classroom learning environments, teacher efficacy beliefs, and the nature of science. She can be contacted at Middle

289

East Technical University, Faculty of Education, Department of Elementary Education, 06531, Ankara, Turkey or jaleus@metu.edu.tr

Randall S. Davies is an assistant professor of educational research at Indiana University South Bend. He graduated from the University of Alberta with a BSc in mathematics and computer science and a BEd in secondary education. He holds a PhD in instructional psychology and technology from Brigham Young University. Dr. Davies taught for many years at both high school and college levels in the field of computer science and IT-related subjects. He has done extensive work developing and evaluating course and instructional materials. Dr. Davies' research interests include methodology for improving teaching and learning and the use of technology in education. He has completed several research and evaluation projects in these areas. He can be contacted at Indiana University South Bend, 1700 Mishawaka Ave., P.O. Box 7111, South Bend, IN, 46634-7111 or radavies@indiana.edu

Tarek Daoud is a teacher at the National Evangelical Institute for Girls and Boys (NEIGB) in Saida, Lebanon where he teaches physics to students following Lebanese and international curricula. He received his License in physics in from the Beirut Arab University and then completed a master's degree in science education at the Department of Education, American University of Beirut. He can be contacted at Department of Education, American University of Beirut, P.O. Box 11-0236, Bliss Street, Beirut, Lebanon or ducitarek@hotmail.com.

Billie Eilam holds a PhD in education from the Technion-Israel Institute of Technology in Haifa. She is currently a senior lecturer in the University of Haifa, Israel, and the director of the Laboratory for Research in Cognitive Processes of Learning. She was the head of the department of Teaching and Teacher Education in the faculty of Education. Her research focuses on the learning sciences and curriculum, with emphasis on visual representations, and the application of learning theories in ecologically valid contexts. Dr. Eilam has authored a book (Hebrew) in addition to chapters and articles in refereed journals (e.g., *The Journal of the Learning Sciences, Social Psychology of Education, American Educational Research Journal, Teachers College Record, Contemporary Educational Psychology, Journal of Research in Science Teaching*, and *Teaching and Teacher Education*). She has presented papers and organized symposia at numerous meetings of professional organizations (e.g., American Educational Research Association (AERA) and European Association for Research in Learning and Instruction (EARLI)). Dr. Eilam served on several national committees in education and won grants for developing curricula and research. Correspondence concerning this chapter should be addressed to Billie Eilam, PhD, Director of the Research Laboratory in Cognitive Processes of Learning, Faculty of Education, University of Haifa, Mount Carmel, Haifa, 31905, Israel. She can be contacted at beilam@construct.haifa.ac.il.

Kirsten Ellenbogen is the director of evaluation and research in learning at the Science Museum of Minnesota. She studies learning in nonschool environments, with emphasis on the role of museums in family life, design of museum learning environments to scaffold science talk, and ways in which scientific visualization technology can engage the public in exploring scientific data and understanding complex phenomena. She was project director, Center for Informal Learning & Schools, King's College London at its inception and affiliated researcher, Museum Learning Collaborative. Dr. Ellenbogen serves on the National Academies Committee on Learning Science in Informal Environments. Publications include articles in *Science Education* and *Environmental Education Research* and chapters in *Cognition in a Digital World, Museum Informatics: People, Information, and Technology in Museums,* and *In Principle, In Practice: Museums as Learning Institutions.* She received her PhD in Science Education from Vanderbilt University and her BA from the University of Chicago. She can be contacted at kellenbogen@smm.org.

Andrea Gay earned her EdD in college teaching of an academic subject-chemistry (CTAS) at Teachers College, Columbia University and completed a postdoctoral research fellowship at the Center for Inquiry in Science Teaching and Learning at Washington University in St. Louis. She is currently an assistant professor of Chemistry Education in the Chemistry and Physics Department at Chicago State University. Her research interests center on means of enabling students to connect chemistry theory with practice in the laboratory and to connect chemistry concepts to their lives. Recent research has focused on incorporating writing and contextualization in laboratory courses to foster chemistry conceptual connections for nonmajors and elementary teachers. Dr. Gay has presented papers and been a reviewer for the annual conferences of the National Association for Research in Science Teaching, the American Chemical Society, and the Association for Science Teacher Education. She can be contacted at agay@csu.edu

Deborah Hanuscin holds a PhD in curriculum and instruction from Indiana University-Bloomington. She currently is jointly appointed as assistant professor of science education and physics at the University of Missouri-Columbia. Her research is focused on elementary science education, and examines epistemological beliefs about science and their role in teaching science. Dr. Hanuscin's work has been published in the *Journal of Research in Science Teaching, Science Education, Journal of Science Teacher Education,* and *Journal of Elementary Science Education.* She has recently presented papers and organized sessions at the annual meetings of several professional organizations including the American Educational Research Association, the National Association for Research in Science Teaching, and the Association for Science Teacher Education. She can be contacted at Professor of Science Education and Physics, MU Science Education Center, University of Missouri, 303 Townsend Hall, Columbia, MO, 65211 or hanuscind@missouri.edu

Colleen New holds a MS in special education from Indiana University South Bend. Currently, she works as a private educational consultant for several local school corporations. She also serves as a research assistant for various faculty projects. Her research interests include assessment of teacher education programs and technology-based interventions for students with cognitive or learning disabilities. She can be contacted at colleennew@sbcglobal.net

Sule Ozkan is a research assistant in the Department of Elementary Education at Middle East Technical University in Ankara, Turkey. Mrs. Ozkan is a doctoral student in the Department of Secondary Science and Mathematics Education in the same university. Her recent research interests include student motivation, epistemological beliefs, learning approaches, self-regulatory learning strategies, and learning environments research. Recently, she has been modeling the relationships among certain learner characteristics and science achievement in her research. She has also conducted studies of the national elementary science curriculum in Turkey. Her elementary Grade 4, 5, and 6 science books were approved by the Ministry of Education to be taught in elementary schools nationally, for 5 years. She is the coauthor of several science activity books at the elementary level. She can be contacted at sozkan@metu.edu.tr

William A. Sandoval is an associate professor in the Graduate School of Education & Information Studies at UCLA. He received his PhD in the learning sciences from Northwestern University. His research focuses on the development of epistemological beliefs about science and the role that instruction plays in mediating such development, and on design-based research methods in education. Dr. Sandoval was a contributing author of the National Research Council report, *America's Lab Report: Investigations in high school science.* He has published and spoken internationally, serves on the editorial boards of the *Journal of the Learning Sciences* and *Science Education*, and reviews for a number of major journals. Dr. Sandoval is a member of the American Educational Research Association, the National Association of Research in Science Teaching, and the International Society of the Learning Sciences. He can be contacted at sandoval@gseis.ucla.edu

Constance Sprague holds an AB degree in zoology and an MS in education, both from Indiana University, Bloomington. She currently is a full-time lecturer at Indiana University South Bend where she teaches classes in methods of teaching science at both the elementary and secondary levels. She is a coauthor of all five editions of *Learning and Assessing Science Process Skills* (Kendall-Hunt Publishing). She supervises practicing teachers in field experiences and student teaching, and has served as an external evaluator on grant projects in science instruction. She can be contacted at Indiana University South Bend, 1700 Mishawaka Ave., PO Box 7111, South Bend, IN, 46634-7111 or csprague@iusb.edu

Dennis W. Sunal holds a PhD in science education, MA in interdisciplinary science, and a BS in physics all from the University of Michigan. He currently is a professor of science education at the University of Alabama. His university teaching experiences include undergraduate and graduate courses in physics, engineering, research in curriculum and instruction, teaching in higher education, and science teaching and learning. He holds both secondary, 6-12, and elementary K-6 teacher certification and has taught extensively on both levels. His research interests are in undergraduate science, preservice teacher education, alternative conceptions and conceptual change in teachers and faculty, and Web course design, pedagogy, and contextual factors in interactive online learning. He has been project director and codirector in numerous grants (e.g., NSF, NASA, Department of Education, USIA, and U.S. Department of Energy). Dr. Sunal has published numerous articles and chapters in refereed journals and books. Recent research presentations have been at the annual meetings of NARST, ASTE, NSTA, SCST, AACTE, and AERA. His published books include *Teaching Elementary and Middle School Science; Integrating Academic Units in the Elementary School Curriculum; Reform in Undergraduate Science Teaching for the 21st Century; and The Impact of State and National Standards on K-12 Science Teaching, and Science in the Secondary School*. He can be contacted at Department of Curriculum and Instruction, Science Education, P.O. Box 870232, The University of Alabama, Tuscaloosa, Alabama, 35487 or dwsunal@bama.ua.edu.

Cheryl White Sundberg holds a PhD in curriculum and instruction and science education from The University of Alabama. She is a retired Alabama State Education officer and currently serves as an adjunct professor at The University of Alabama. Her research interests are science education, technology, and distance learning. Dr. Sundberg has coauthored three chapters and five refereed articles in several journals. She has recently presented papers at the annual meetings of professional organizations including the American Educational Research Association, the National Association for Research in Science Teaching, National Science Teachers Association, and Association for Science Teacher Educators. Among her other professional activities is her role as editor of the newsletter for the Council for Elementary Science International. She can be contacted at sundbergrc@att.net

Ceren Tekkaya is associate professor of elementary education programs at the Middle East Technical University in Ankara, Turkey. She completed her doctorate at Middle East Technical University in 1996. Dr. Tekkaya's research interests include elementary and secondary school students' conceptual understanding of science and teaching methods. She can be contacted at Department of Elementary Education, Middle East Technical University, P. O. Box 06531, Cankaya, Ankara, Turkey or ceren@metu.edu.tr

David A. Ucko serves as deputy director for the Division of Research on Learning in Formal and Informal Settings at the National Science Foundation. Previously, he was section head for science literacy and program director for informal science education. He also is president of Museums+*more* LLC. Formerly, he served as executive director, Koshland Science Museum, National Academy of Sciences; founding president, Science City at Union Station; president, Kansas City Museum; chief deputy director, California Museum of Science & Industry, Los Angeles; and vice president for programs, Museum of Science & Industry, Chicago. Dr. Ucko was a presidential appointee with senate confirmation to the National Museum Services Board. He taught at the City University of New York and Antioch College, where he wrote two college chemistry textbooks. Ucko is a AAAS Fellow and a Woodrow Wilson Fellow who received his PhD in inorganic chemistry from MIT. and BA from Columbia. He can be contacted at DUcko@nsf.gov

Craig A. Wilson earned his PhD in curriculum and instruction from the University of Toledo. While working on his graduate degree, he was presented with the Robert R. Buell Award for Excellence in the Sciences and Humanities. Dr. Wilson is an associate professor of education in the

Early Childhood and Elementary Education Department at East Stroudsburg University in Pennsylvania, where he teaches undergraduate and graduate science methods courses and a graduate research course. His primary research interests are in field-based experiences, inquiry-oriented instruction, and technology integration. He has recently been published in *Science Scope* and *Best Products in Mathematics and Science*, a monograph from the collaborative for excellence in teacher preparation in Pennsylvania. He is the project leader for a grant program at East Stroudsburg University, titled Middle School Science Certification for Special Education Preservice Teachers. Supported by the Pennsylvania Department of Education, this project has brought four science professors and three education professors together to work on a highly collaborative program. He can be contacted at CWilson@po-box.esu.edu.

Emmett L. Wright, PhD is a professor of science/environmental education at Kansas State University. In recent years he served as the served as the director of research for the National Commission on Mathematics and Science Teaching for the 21st Century, and as a program director and TPC section head in the division of elementary, secondary education and informal education at the National Science Foundation. He was awarded a Fulbright Scholars position at the University of Malta for the 2006-07 academic year. He holds a PhD from Pennsylvania State University (science education/environmental biology). Dr. Wright's research interests include decision-making attitudes, problem solving, misconceptions, and scientific discrepant events, and international education. He has over 130 publications and has served as president of NARST and on the board of directors of NSTA and SCST. He has received major curriculum-development, teacher-education, and research grants, from EPA, NSF, U.S. Department of Education, U.S. Department of State, and U.S. Department of Energy. He can be contacted at birdhunt@ksu.edu

Printed in the United States
132352LV00002B/1-10/A